A BALL, A DOG, AND A MONKEY

1957—The Space Race Begins

MICHAEL D'ANTONIO

SIMON & SCHUSTER PAPERBACKS New York London Toronto Sydney

SIMON & SCHUSTER PAPERBACKS
A Division of Simon & Schuster, Inc.
1230 Avenue of the Americas
New York, NY 10020

First Simon & Schuster trade paperback edition September 2008

SIMON & SCHUSTER PAPERBACKS and colophon
are registered trademarks of Simon & Schuster, Inc.

For information about special discounts for bulk purchases,
please contact Simon & Schuster Special Sales at
1-800-456-6798 or business@simonandschuster.com

Designed by Jaime Putorti

Manufactured in the United States of America

10 9 8 7 6 5 4 3 2

The Library of Congress has cataloged the hardcover edition as follows:

D'Antonio, Michael
  A ball, a dog, and a monkey—1957, the space race begins / Michael D'Antonio.
     p. cm.
Includes bibliographical references and index.
1. Artificial satellites—United States—History. 2. Artificial satellites—Soviet Union—
History. 3. Space race. I. Title.

TL796.5.U5 B54 2004
629.43'409045—dc22
                                    2007017619

ISBN-13: 978-0-7432-9431-7
ISBN-10:   0-7432-9431-9
ISBN-13: 978-0-7432-9432-4 (pbk)
ISBN-10:   0-7432-9432-7 (pbk)

Praise for *A Ball, a Dog, and a Monkey*

"A Best Science Book of 2007"   —NPR's *Science Friday*

"[D'Antonio's] fast-paced narrative incorporates first-hand accounts of every-day citizens caught up in the excitement of America's push into space."
—*Publishers Weekly*

"D'Antonio skillfully captures both the energy and the urgency of the U.S. drive toward space. His narrative is further enlivened by firsthand accounts from U.S. citizens recalling their excitement during those heady days."
—Marjorie Kehe, *The Christian Science Monitor*

"A fun, fast read. . . . Amazing we had to wait 'til the 50th anniversary of that year's tumultuous events to get the popular history they deserved."
—Angela Gunn, USAToday.com

"Pulitzer Prize–winning writer Michael D'Antonio tells the story of the first year of the space race, sparing none of its panic, scientific fervor and zaniness. . . . D'Antonio's narrative is brilliant and, best of all, extremely entertaining."   —Larry Cox, *Tucson Citizen*

"Entertaining. . . . D'Antonio delivers the technological heroics on which spaceflight fans are keen."   —*Booklist*

"Fascinating and fact-filled. . . . [D'Antonio] offers enough nuclear-war anxiety, launch-pad countdowns and orbiting mammals to turn an idle evening of reading into a nail-biting experience."
—Doug Childers, *The Richmond Times-Dispatch*

"If you remember going out into the backyard on nights and craning your neck to see that pinpoint of moving light that was Sputnik, you owe it to yourself to read this book."   —Dan Danbom, *Rocky Mountain News*

"A genial look at the earliest days of the space race. . . . The story's charm lies in D'Antonio's evocation of the average American's response to the dawning space age. . . . Recovers for a new generation the thrill of a pioneer quest and the spirit of an age that already seems like ancient history."   —*Kirkus Reviews*

"D'Antonio . . . brings a human perspective to the Sputnik story."
—*Library Journal*

ALSO BY MICHAEL D'ANTONIO

Hershey: Milton S. Hershey's Extraordinary Life of Wealth, Empire, and Utopian Dreams

The State Boys Rebellion

Tour '72: Nicklaus, Palmer, Player, and Trevino: The Story of One Great Season

Mosquito: The Story of Man's Deadliest Foe (with Andrew Spielman)

Tin Cup Dreams: A Long Shot Makes It on the PGA Tour

Atomic Harvest: Hanford and the Lethal Toll of America's Nuclear Arsenal

Heaven on Earth: Dispatches from America's Spiritual Frontier

Fall from Grace: The Failed Crusade of the Christian Right

★

FOR TONI, ELIZABETH, AND AMY, AS EVER

CONTENTS

PRELUDE: Ford Fairlanes, Atom Bombs, and Satellites   1

ONE: Those Damn Bastards!   17

TWO: New Moon Worries   39

THREE: The Cape   65

FOUR: And a Dog Shall Lead Them   85

FIVE: Hold It!   105

SIX: The Acid Test   129

SEVEN: Being Nonchalant and Lighting Up a Marijuana   149

EIGHT: A New Era of Exploration   171

NINE: Opportunists and Adventurers   191

TEN: Eggheads and Pie Trucks   209

ELEVEN: The Monkey and the President   227

Epilogue   249

Acknowledgments   261

Notes   263

Index   293

A BALL, A DOG, AND A MONKEY

Ford Fairlanes, Atom Bombs, and Satellites

In September 1957 a privileged and adventurous American family began a European road trip that would take them from the Bug River at the Polish border, where a sign dedicating the highway read "FOR PEACE IN THE WHOLE WORLD," to Moscow, roughly 600 miles to the east. After twenty years, the Union of Soviet Socialist Republics had opened to tourists the route previously followed to defeat by Napoleon's Grande Armée and, more than a century later, by Hitler's tanks. Among the first to be admitted, reporter James "Scotty" Reston had packed his wife, Sally, and his nearly grown son, Richard, into a brand-new car for a journey that would yield a front-page story for readers of *The New York Times*.

"Want to drive to Moscow?" began Reston's account, which was published on October 1, 1957. "You can now do it with the blessing (and supervision) of the Soviet Government if you have the money and the patience."

Although Reston's article promised something new—American innocents on the road behind the Iron Curtain—it was mostly a familiar and reassuring report on the inferiority of communism, Soviet society, and its products. Though always kind to the Soviet people, and reflexively fair, Reston recognized the shortcomings of a system that discouraged enterprise and controlled everything from ideas to industry. This truth was evident in almost every paragraph.

In Byelorussia, where the landscape was all peat bogs and dilapidated farms, Reston saw so few modern vehicles that peasants could lay their sheaves of wheat on the new highway to dry. (When requested, he obligingly drove his car back and forth over the stalks to aid the threshing.) In Minsk, reported Reston, the food was bland, the beer watery, and the wine almost vinegar. In Smolensk the roads were choked with mud and the houses were hovels. Across the country the hotels were overpriced and underfurnished. At 54 octane, Soviet gas—called benzene—was a poor diet for an American V8.[1]

In the 1950s, dilapidated cities, shortages, and mediocre goods were facts of life that set the Soviet nation apart from the West and, most of all, from the United States, where the postwar period had brought a never-ending geyser of "new." In America, where the famous Levittown had just turned ten years old, suburban development offered new levels of comfort for growing numbers of families who could be counted as solidly "middle class." Novel technologies and sources of wealth, from atomic energy to plastics, were the subject of near constant conversation in the press. Opinion polls showed that the country had never been happier. And it was widely understood that barring an awful sky-darkening catastrophe—a possibility that everyone struggled to ignore as they went about their shopping—tomorrow would always be better.

The catastrophe that everyone imagined, of course, would be a nuclear war with the very same Soviets who had so much trouble making decent beer and gasoline. Despite the ravages of World War II (20 million dead, and an economy in ruins) the USSR spent billions of rubles annually to compete with America in developing weapons. An early sign of their success came in the skies over Korea in 1950 when U.S. fliers encountered swept-wing MiG-15 fighters that were surprisingly effective opponents. Next, in 1953, the Soviets shocked Western experts, who predicted they were at least four years away from such a feat, by exploding a hydrogen bomb. In this moment, America's overwhelming nuclear superiority, represented by the poisonous mushroom clouds roiling above the deserts of the West after test shots, seemed to evaporate. Suddenly both sides could deliver the equivalent of one million tons of TNT—one-third of *all* the firepower used in World War II—with a single bomb.

After World War II, as the only two states armed with atomic weapons and the means to deliver them, the Soviet Union and the United States of America occupied similarly nervous psychological and strategic positions. They maintained a constant state of stalemate, planning attacks while knowing that any first move would bring massive retaliation resulting in the deaths of perhaps millions of civilians. This state of affairs, known later know by the acronym MAD—Mutual Assured Destruction—was inherited by President Dwight Eisenhower when he took office in 1953. At his inauguration he also inherited Strategic Air Command chief Curtis LeMay, the Air Force general who oversaw the bulk of the nation's atomic arsenal. LeMay was the walking, talking, cigar-chomping embodiment of MAD, which he preferred to call "Pax Atomica."

There was no small irony in Ike's commitment to this policy and to LeMay's stewardship. As the Allied supreme commander in Europe during World War II, Eisenhower had initially opposed the massive air assaults on German cities that had been advocated by General LeMay and others. Eisenhower had also stood against the atomic attacks on Japan, arguing that they were unnecessary from a military standpoint and risky where world opinion was concerned. But by the time he was president, the logic of maintaining the threat of atomic cataclysm in order to keep the peace was so central to the superpowers' relationship that Eisenhower had no choice but to continue it.

With such a nerve-wracking reality came much fear, anxiety, and even paranoia. The Soviets scanned American military bases, which were dotted around the world, and saw threats in every direction. "Our country," explained one Soviet official, "was literally a great big target for American bombers operating from airfields in Norway, Germany, Italy, South Korea and Japan."

For their part, American defense leaders regarded the USSR to be, in the words of one national security assessment, "inescapably militant" and devoted to the destruction of its main adversary, the United States. When they looked at the globe, they saw how easy it would be for Soviet aircraft and missiles to come over the Arctic and strike a score of major cities. They counseled both vigilance and constant improvements in American military capabilities.

Of course vigilance has its side effects, including anxiety and fear, which were reflected in the art and literature of the day, including a host of atomic disaster novels. The most notable, *On the Beach* by Nevil Shute, takes place in Australia after the rest of the world has been blown to smithereens and everyone left alive was destined to die of radiation sickness. Published in early 1957, it became an immediate international bestseller.

While Shute played with nuclear terror as an artist, America's big defense contractors exploited it for business purposes. Companies that made aircraft, ships, rockets, and electronics spent heavily on advertisements in national magazines to tell taxpayers about the value of their work on weapons systems. Often these companies touted expensive new technologies that were only under consideration in Washington, hoping to influence decisions on funding. In 1957, North American Aviation, for example, bought two-page spreads in *Time* magazine to ballyhoo a 2,000-miles-per-hour bomber that was never added to the U.S. force.

Although this particular ad didn't pay off, defense executives put money into big media promotions because politics played a big role in government decisions about weapons systems. Indeed, contractors were able to compete for huge contracts because national defense was a popular priority. On both sides of the Cold War, the public supported expenditures on defense because they felt both afraid and morally superior. The man on the street, whether in Moscow or New York, reflexively searched for signs that his side was better.

For three weeks in July 1957, New York area commuters who needed an extra shot of encouragement could gaze at the sixty-three-foot Redstone rocket on display beneath the stars painted on the vaulted ceiling of Grand Central Terminal. Created with the help of former Nazi

V-2 rocket engineers who had been brought to America after World War II, the Redstone was designed to throw both conventional and nuclear warheads hundreds of miles. It was hauled to the terminal and installed by its manufacturer, the Chrysler Corporation.

In general, the world's richest economy—symbolized by the truly grand train station and its powerful military—gave Americans reason to feel reassured. But behind their confidence lurked an insecurity that was new to a land that had long felt blessed, by two great oceans, with strategic invulnerability. Everyone understood there was no defense against missiles armed with atomic bombs, and it was hard to feel secure in a peace that depended on a precarious balance of terror. Worst of all, from the American perspective, the Soviet bombs were controlled by Nikita Khrushchev. Khrushchev was widely believed to be emotional, shrewd, unreliable, and dangerous. Typical of Western reporting, journalist Stewart Alsop, writing in *The Saturday Evening Post*, would describe Khrushchev as a charming but "passionately ambitious, carefully calculating, bloodstained man."

Cousins of Franklin Roosevelt, the influential writers Stewart and Joseph Alsop were fervent anticommunists who feared the Soviet military buildup and favored the strongest responses to signs of aggression. Since brother Joe had been rebuffed when he tried to make a deal to get inside information from the White House, the Alsops were not inclined to cut Eisenhower any slack. In the summer of 1957, Stewart Alsop had reported Soviet gains in rocketry in grave terms. His articles contributed to public anxiety and annoyed both Pentagon officials and diplomats, who considered the brothers alarmists. Eisenhower reportedly "blew his stack" when Stewart also revealed cuts in funding for U.S. weapons programs and wrote that the president risked losing the race to produce an intercontinental ballistic missile in order to cut taxes before the coming midterm election.

With the Alsops and others drumming on their Cold War nerves, the American public hungered for reassurance. This desire helps to explain why so many people eagerly consumed press stories that highlighted Russian shortcomings and confirmed their own country's superiority. According to these reports the communists walked through a life that was cold and gray as the steppes while Americans sped down modern highways in futuristic cars. This reality was confirmed for Richard Reston on his family trip. Years later he would retain only fuzzy memories of the Soviet countryside and the people. But he had vivid recollections of the car.[2]

★

The car was an automotive wonder called a Ford Fairlane (named after the pioneering capitalist Henry Ford's estate, Fair Lane) that had been shipped new from its factory to Europe so the Restons could make their drive. Decorated with heavy chrome bumpers and matching grille, the Fairlane was a marvel of excess. Buyers could choose from a long list of options, including thirty-seven different fabric combinations and a custom tissue dispenser. The Restons' Fairlane was painted in graceful swaths of black and orangey-gold that evoked a Baltimore oriole in flight. With its big engine, a length of 210 inches, and a wheelbase just shy of ten feet, the Fairlane was massive and powerful. Most remarkable of all, it had a retractable "Skyliner" hardtop roof. With the push of a button a thirty-second mechanical ballet began. Seven different motors worked together to open the trunk, fold up the panels of the roof, shove them inside, and then close the lid.

The hardtop ballet was performed at almost every stop on the Restons' drive to Moscow, often at the request of police officers, who pushed the button themselves. Wherever they parked the Americans were quickly surrounded by men, children, and women in babushkas who admired the Fairlane and asked them about its power, speed, and price. (The car cost $2,945, more than half the average annual wage of $5,500 in America.) Crowds gathered around the car even when it was unattended. Many of the gawkers took photos. Other caressed it. One day the Restons returned to where the Fairlane was parked to discover an artist making a sketch.

Scotty Reston, a man partial to curved pipes, bow ties, and jaunty hats, usually handled the demonstrations of the retractable hardtop, ceremoniously pressing the button and then sitting calmly as the Fairlane performed. However, on a day when the car was parked outside the National Hotel and his father was at the Kremlin for an interview, the then twenty-year-old Richard assumed this patriotic duty. With a crowd gathered 'round he climbed into the driver's seat and hit the switch. The trunk lid opened to receive the hardtop, but then, for the first and only time on this trip, the mechanism jammed halfway through its routine.

"Here was the smart-aleck kid showing off the technological sensation from the West and he messes it up," recalled Reston, decades later. "I'm sure people passing by thought to themselves 'Who are these idiots trying to get that car roof unstuck?'"

After a brief struggle, the roof was freed and then folded neatly away, but not before young Mr. Reston had felt a brief shudder of vulnerability. Western technology was not perfect. The rich and the powerful could be embarrassed.[3]

★

While the Reston family was motoring in splendor around Moscow, a determined Soviet engineer arrived in the capital with a historic purpose. Short and heavy, with deep-set brown eyes ringed by dark circles, this fifty-one year-old engineer and bureaucrat had devoted his entire adult life to achieving a fantastic goal, which was at last within his grasp. All he needed now was permission from a panel of overseers gathered in a group called, the State Commission.[4]

Although soon his work would be hailed around the world, the engineer's name would remain a mystery for years. For security reasons, and so that the state would receive credit for his achievements, he would be referred to only as the Chief Designer.

To reach this moment, the Chief Designer had suffered in ways that were possible only in a totalitarian world. In 1938 he had been caught up in one of many purges carried out by Stalin, who in his paranoid reign jailed roughly 5 million of his countrymen and killed about a million on suspicions of treason. Tortured and beaten until he confessed to trumped-up charges, the Chief Designer spent six years in various prisons, including the notorious Kolmya Arctic camp, where the annual death rate was 30 percent. Along the way he came close to starving to death and nearly died from exposure.[5]

Having survived the gulag, the Chief Designer was "rehabilitated" after he completed studies in Marxism-Leninism at an evening univer-

sity. In the spring of 1957 the Soviet government officially acknowledged that he had been wrongly imprisoned. By then he had recovered
so thoroughly, both professionally and personally, that he seemed almost
unaffected by his imprisonment. Like many purge survivors, even his
loyalty to the communist way and to Stalin were unshaken. Only those
who knew him well recognized the subtle signs—like his tendency to
collect and consume the crumbs around his plate before he left a
table—that betrayed his past suffering.

This resilience was a product of the Chief Designer's tempera-
ment and childhood experience. Raised without a father by a mother
who was often absent, he was a self-directed and capable boy. As an
adult he exuded both personal confidence and a pragmatically dismal
outlook on history. The Chief Designer was certain that he was destined for greatness, but would often say that soon "we will all vanish
without a trace."[6]

In his pessimism the Chief Designer was imagining, perhaps, one
possible outcome of his work on behalf of the Soviet military. Obsessed
with flight, he had helped lead the development of a series of rockets
that culminated with the R-7, a giant capable of hurling one of his
country's bulky atomic bombs a distance of 4,350 miles. Progress toward
the R-7, which had five engines, had not been perfect. The first one
launched came apart after one minute of flight and crashed. The second
didn't make it even to the one-minute mark.

Finally in August of 1957, a month before the Restons drove to
Moscow, an R-7 rocket flew for more than 4,000 miles. On a personal level, the success showed that the Chief Designer possessed
the skill to design such a rocket and the strength of will to produce
it, despite the brutish Soviet bureaucracy. For the world, the flight of
the rocket, which the Soviet crew called "Old Number Seven," had a
more ominous meaning. As the first launch vehicle capable of carrying a nuclear weapon from one hemisphere to another, the R-7 put
all of Europe and parts of North America within range of the USSR's
nuclear arsenal.

Military officials in the United States knew enough about the R-7
to consider it unmatched in range and power. Intelligence sources, including radar installations in Turkey, confirmed the Soviet claims for

its successful test. The more hawkish factions in the U.S. Defense Department immediately saw military superiority shifting to the East and complained about budget restrictions and policy failures that seemed to be holding back the development of a comparable American missile. Others, especially leaders outside the military establishment, urged calm and diplomacy. Secretary of State John Foster Dulles, for example, responded to the rocket's flight by repeating President Eisenhower's previous call for an end to the race toward new "appalling means of human destruction."[7]

Worried as they were about atomic attack, America's generals and civilian defense officials were less focused on the other purpose the Chief Designer imagined for his invention, the conquest of space. However, space *was* the real goal he had held in mind for most of his life. It was this obsession that had helped him to endure the gulag, the frustrations of the Soviet system, and the physical hardships at the USSR's isolated, secret rocket facilities. If he could reach outer space, the sacrifices would be justified.

The sacrifices were many. The main Soviet launch center was located on a steppe where the winters were so severe that a man who lost his way in a storm would literally freeze to death. It happened to horses too. Not surprisingly, morale was a problem. Once each week, typically on Thursday, the Chief Designer spent hour after hour listening to the complaints and requests of underlings desperate for a solution to their problems. Sometimes the issue was related to the work at hand, but he might also have to resolve problems related to health care or housing. On Reception Day, as it was called, almost any request imaginable might arise, and like a good king (or department store Santa) receiving supplicants, he was expected to hear each one and offer an effective response.

Whatever they were, the problems presented by workers and colleagues on Reception Day were much less difficult to handle than the task the Chief Designer faced getting high-quality work out of the Soviet industrial system. To put an object into space he would need materials never before made in his country, revolutionary electronics, and machine work done to exquisite tolerances. To get access to funds and scarce facilities—the Soviets lagged far behind the West in many industries—the Chief Designer would have to compete with many other high-priority projects. For example, a single-stage nuclear-tipped missile called the R-12 was being tested at a second Soviet rocket base called Kapustin Yar, which was on the banks of the Volga River in Soviet Russia. Officials were also investing in a submarine capable of launching R-11 nuclear-tipped missiles, improved nuclear bombs, and defensive radar and missile systems.[8]

Compromise was unavoidable. Unable to produce the lighter materials and smaller-scale payloads envisioned by American rocketeers, the Soviets were forced to build a much more powerful rocket, so it could lift extremely heavy weights. (Nearly 100 feet tall, the R-7 weighed 100 tons. The Soviet H-bomb weighed five.) Further compromise was required when the industrial bureaucracy fell behind in constructing the sophisticated satellite instruments envisioned for the first mission to space. The Chief Designer had to quickly imagine and then build (without detailed drawings) a much simpler setup that included radio equipment, batteries, and a cooling system, which would be sealed inside a thirty-pound metal ball that was less than two feet in diameter. All this to achieve the simple, nonmilitary goal of sending into orbit an object that could be monitored from Earth.

But here he reached the end of all possible compromises. To protect the instruments inside from heat, the ball would have to be extremely reflective, so the sun's warming rays would bounce away. No flaws, smudges, or inconsistencies could be tolerated. It was difficult to wring such exacting work out of the Soviet system. Worried that shop workers wouldn't grasp the seriousness of their task, the Chief Designer supervised them personally. To get his point across, he put his satellite on a special stand, wrapped it in velvet, and announced that this model of satellite should be made with the care that would go into an artist's masterwork. "This ball," he declared, "will be exhibited in museums!"

By mid-September 1957 the first shiny ball was done and workers had attached to it four long antennae that jutted out in the same direction, giving the satellite the appearance of a long-legged spider. When loaded with gear (the batteries were the heaviest part) the satellite weighed about 180 pounds. It was formally named *Prosteshyy Sputnik*.

The name, which meant "simplest fellow traveler," would soon be short-ened to simply *Sputnik*.

In Moscow, as he pressed for permission to launch his shiny ball, the Chief Designer warned his superiors that the United States was also driving toward space, a fact that was well established by press reports. According to legend, his anxiety about America's satellite program had been inflamed by a clerical error. Apparently a communiqué about a scheduled conference on satellites had been so poorly transcribed it seemed to say that the United States was planning the imminent launch of a satellite, rather than a meeting to talk about one, which was the actual case.

Thus motivated, the Chief Designer declared that his own satellite was built and ready—a surprise to the State Commission—and pointed to August's successful R-7 launch as proof that his team could get it into space. When the commission resisted, he cornered them by suggesting they hand the decision off to the Presidium of the Commu-nist Party. Unwilling to be second-guessed by this higher authority, the commission relented and gave him permission to launch the ball on October 6.[9]

★

On September 29, 1957, two days before Scotty Reston's Poland-to-Moscow travel piece would be published on the front page of *The New York Times*, the Chief Designer boarded an airplane in Moscow for a 1,300-mile flight to the barren steppe of Kazakhstan and the USSR's secret rocket launching complex.

Besides putting it in a place that could only be described as godfor-saken, the Soviets had gone to great lengths to hide the largest of its *cos-modromes*—it covered 660 square miles—from the outside world. Its name, Baykonur, was a lie. The real Baykonur was a mining town that was 200 miles away. The launch site was actually twenty miles north of a place called Tyuratam, which itself was little more than a stop on the railroad connecting Moscow and Tashkent. (A spur ran north from Ty-uratam to the launch site.)

Although the terrain was ideal for tracking missiles, the treeless land-scape punished its inhabitants with swirling dust, searing summer heat, and brutal winter cold. In warm weather, disease-carrying mosquitoes swarmed in huge numbers. In cold weather, rats that spread bubonic plague sought shelter in walls and crawl spaces. It was such a forbidding environment that the tsars had used it as a place of exile for, among others, the nineteenth-century artisan Nikifor Nikitin, who was sent there for promoting the idea that men could fly to the Moon. Later it was the site of open pit copper mines, which may have been worked by political prisoners in the 1930s. In the 1950s conditions for miners and others in the *real* town of Baykonur were far better, especially during the time when locals were able to trick Soviet bureaucrats, who thought they were dealing with the space center, into sending them huge shipments of supplies, including precious cement and timber for construction.[10]

Fortunately for the Chief Designer, when he reached the launch site his rank earned him quarters inside one of the few small houses, fenced and flanked by scrawny trees, that had been built on the windy steppe. Not that he spent much time there. Worried about his American rivals, he bore down hard on his team. After supervising the rocket's assembly in a large aircraft hangar, the Chief Designer noted the process was going so smoothly that the launch could be advanced by two days. Eager to be the first in space, he contacted Moscow and received permission to accelerate the schedule.

On October 3 all was ready at Baykonur. The Chief Designer led a group of other engineers as they walked in front of the rocket, like a procession of priests, as it was moved by locomotive along a short railroad track from its hangar to the launch pad about a mile away. Once the R-7 was set in its place, huge metal struts that angled up from the pad to the rocket's flanks were raised and secured. With bright floodlights illuminating the pad, work continued through the night. Just before 6 A.M. on the morning of October 4, crews began to load the machine with liquid oxygen and kerosene that, when combined and ignited, would provide 876,000 pounds of thrust.

The launch team worked through the day, testing various systems on the ground and in the rocket. Just one significant problem arose. An oxygen leak became the subject of a heated argument—the Chief Designer lost his temper—but the problem was quickly resolved.

Night fell, the countdown continued. The steel supports were drawn back from the rocket. In the blockhouse where he would observe the launch, the Chief Engineer was crowded by members of the State Commission, who had flown in for this moment. Their excitement and anxiety grew as the clock ticked toward liftoff.

As he watched and waited for signs that would force him to halt the countdown, the Chief Designer could only trust that his decades of sacrifice, devotion, suffering, and obsession had been enough. These were powerful factors that should insure success, but according to legend, and reliable sources, he had also performed a good-luck ritual that had become a tradition for Soviet rocket men. Hours before, when it had been safe to approach, he had walked out to stand beside his towering creation, opened his pants, and pissed on it.[11]

At 10:23 P.M. Moscow time, a light flashed on a monitoring panel in the Baykonur control room, suggesting a minor fuel problem. The Chief Designer chose to ignore it. There were no more glitches, no malfunctions to force him to stop.

At 10:28, when the countdown reached zero, a young lieutenant named Boris Chekunov pressed the button to ignite the R-7. Fire and smoke poured out of the roaring engine as the massive rocket lifted off slowly and then picked up speed. After the rocket was away, radio operators tracking its performance soon reported it was perfectly on course. Racing into the night sky at roughly 18,000 miles per hour, it carried *Sputnik* on a roaring flame that could be seen by precious few as it arced over scattered farms and a vast desert, racing toward Siberia.

Facilities at Baykonur were so limited that there was no shelter big enough to accommodate the dozens of engineers, technicians, and observers who witnessed the launch. For this reason many had to stand together outside, waiting for a report from a tracking station on the Kamchatka Peninsula, more than 3,000 miles to the east. Soon word came that this listening post on the edge of the Pacific had picked up the distinctive beep of the satellite's radio transmitter. Excitement swept through the crowd, but the Chief Designer refused to celebrate. He would not declare the mission a success until his creation was heard as it returned from the other side of the world.

Roughly an hour after the report from Kamchatka, one listening post and then another reported hearing *Sputnik*'s return to the sky over the USSR. Tears filled the Chief Designer's eyes as his dream was achieved. "Congratulations," he told the crowd at Baykonur, "the road to the stars is now open." After one more orbit, one of his colleagues called to inform Khrushchev, who was in Kiev for meetings at the Marinsky Palace.

The chairman congratulated the men at Baykonur and returned to his conferences. Reports on Khrushchev's demeanor that night vary. Some said he was beaming as he made a brief announcement about the satellite. "The Americans have proclaimed to the world that they are preparing to launch a satellite of the earth," he said, according to one source. "Theirs is only the size of an orange. We, on the other hand, have kept quiet. But now have a satellite circling the planet. And not a little one but one that weighs eighty kilos."

Others say that Khrushchev didn't immediately grasp the significance of the accomplishment and its propaganda value. According to this view, the Soviet leader prized only advances with apparent practical purpose, like new tractors and airplanes. A year before he had used an untested Tu-104 jet to send embassy mail to London, just so he could show off the second jet-powered airliner ever flown. (Britain's de Havilland Comet was the first.) He grasped the value of a fast new plane, but the full meaning of *Sputnik* escaped him.

Certainly the initial Soviet announcement confirmed the idea that Khrushchev was lukewarm. The party newspaper *Pravda* published a mere three paragraphs on the event, supplied by the official news agency, Tass, and the item did not lead the front page. The article credited "a large, dedicated effort by scientific research institutes and construction bureaus" but revealed not a single name associated with the satellite. It said nothing at all about the military or political implications of this achievement.

It would be up to the rest of the world to signal the importance of what the Chief Designer had accomplished. Soon enough, it would be obvious to all that he had, in one moment, opened a new frontier for humanity, changed the balance of the Cold War competition, and made the Ford Fairlane irrelevant. Richard Reston, at school in England when the news came, would remember, most vividly, how classmates suddenly seemed concerned about America's strength and status in a dangerous historical moment.

"They considered it a great national failure for the United States," he would recall. "It was a shock. People admired what the Soviet Union had accomplished, and were struck by how they had outdone the Americans. We [American students abroad] were a bit defensive about it. We had been taught that America had all the answers and suddenly we didn't."[12]

ONE

Those Damn Bastards!

On October 4, 1957, the USS *Glacier*, stubby and wide-beamed, chugged westward in the Pacific Ocean, tracing a course a few hundred miles north of the Galápagos Islands. Shaped like a watermelon cut lengthwise, the ship was designed to crush through polar ice. Her engines were powerful and her steel skin was thick enough to survive even high-speed collisions with growlers, small bergs that sounded like angry dogs as they scraped along the hull. On her maiden voyage in 1955—called Operation Deep Freeze—the *Glacier* had cut through 400 miles of frozen ocean that would have stopped any other American ship.

But as sturdy as she was, in the open sea the *Glacier* was a bit tipsy, rolling with almost every wave, which complicated matters for civilian physicist James Van Allen, his student Larry Cahill, and the members of the crew who gathered early in the morning to help launch a strange contraption called a "rockoon." First they hauled a huge, clear plastic balloon—source of the "oon" part of rockoon—out of the ship's hold and onto the quarterdeck, which was otherwise used as a helipad. Next came bottles of helium, which were tapped to fill the balloon with 26,000 pounds of gas. As it grew larger, and was subject to the breeze, the balloon flopped around like a fish on the deck.

As the crew continued the task of inflating the balloon, Van Allen turned to a long, narrow wooden crate that was propped on sawhorses. Inside was an Aerobee rocket—the "rock" in rockoon—that was roughly six feet long and not much bigger around than a rolling pin. First developed in the late 1940s, various generations of the Aerobee would serve for decades as reliable rockets for exploring the atmosphere and near space. As he turned to this particular rocket, Van Allen summoned some helpers and began to review what they had to do to make it airworthy.

A slightly built man of forty-three, with sloping shoulders and a soft, boyish face, Professor Van Allen hardly looked like a commanding figure. But he had spent more time at sea than most of the crew, and he directed them with a combination of certainty and respect that inspired cooperation. With help from Cahill he got the rocket out of its box. They tested its radio transmitter and attached a rope that would connect it to the balloon for a journey aloft.

The rocket was the tricky part. One had recently fired, accidentally, while it was being readied for use on the deck of another ship. Ignited by a signal from its own radio transmitter, the thing took off horizontally. The flame from its tail burned a ship's officer who stood nearby and blew the coat right off one of Van Allen's students from the University of Iowa Physics Department. It then smashed through a pair of sawhorses and headed straight for a sailor who was talking to the bridge via a deck phone. The rocket sliced through the cord on the phone, leaving the shocked crewman unscathed as he held the suddenly un-tethered receiver in his hand. The rocket finally crashed into a stack of helium canisters that were, to the relief of everyone on board, empty and not pressurized. Burning fuel and pieces of metal flew all over the deck.

Although the injuries from that accident included just the burns suffered by the officer and damage to the student's eardrums, it reminded every scientist, officer, and enlisted man involved with the rockoons that while they looked like mere gas bags trailing fireworks, they were potentially lethal.

Aboard the *Glacier*, as the moment for the October 4 launch approached, Lieutenant Stephen Wilson picked up a phone on the deck and contacted the ship's bridge. The balloon could be released only when the ship was traveling with the prevailing breeze, at exactly the same speed as the wind, so that the entire contraption would clear the *Glacier*'s smokestack, antennae, and mast. To get the course and speed right, Wilson barked a stream of orders to a helmsman, who made adjustments. "If anyone looked at your ship's track when we were doing this," he would one day recall, "they would be convinced there was some drunk at the helm."

At 1:16 P.M., with Wilson keeping the ship steady, Van Allen signaled the men who held the balloon to let it go. The big clear bubble of gas rose, pulling behind it the rope and rocket and leaving the ship behind. Looking much like a jellyfish with a single tentacle, the balloon moved slowly at first, but as it climbed, and the atmosphere thinned, it picked up speed. Below, Van Allen watched through binoculars as the sunlit blob of plastic flew farther and farther away. Beside him a radioman, Petty Officer David Armbrust, stood wearing earphones and holding a special antenna that was made to track signals from the rocket.

As it approached an altitude of fifteen miles, where the rocket's engine would automatically fire, the men below had one last chance to worry about whether the rig had somehow gotten tangled. A rocket pointing in the wrong direction could race back down to Earth, and hit the ship like a bomb. (This had never happened, but you never know.)

Fortunately, the rockoon performed as planned. At the fifteen-mile mark it fired properly, breaking free from the balloon and climbing skyward to deliver its payload—nine pounds of scientific instruments and a radio transmitter—to a final altitude of seventy miles. Below, on the deck of the ship, Van Allen directed Petty Officer Armbrust, who struggled to keep his antenna pointed at the right spot in the sky so he could pick up the rocket's signals.

When it reached its apogee the rocket was beyond the stratosphere and on the edge of outer space. The equipment on board measured the amount of radiation streaming to Earth from various sources in the universe and radioed the data back to the *Glacier*. With his rockoons Van Allen had already pierced other regions of the sky. His goal was to

understand the earth's magnetic field and the quality, quantity, and behavior of cosmic rays throughout the atmosphere. Because these phenomena affected radio signals, this work had some immediate practical
applications. But it would also provide vital information about the
conditions a rocket, satellite, or human being might one day encounter in space.[1]

Though connected with Van Allen's long-term research, this particular rockoon flight was part of an ongoing, international campaign of exploration that involved 60,000 people in nearly seventy countries and
was called the International Geophysical Year. The IGY had been con-
ceived seven years earlier in Van Allen's living room by a group of renowned scientists who fed on his wife's homemade seven-layer cake. It
was modeled after two previous "Polar Years" held in 1882 and 1932,
which also involved the world's top explorers and scientists in similar efforts to study the Earth's condition, from pole to pole, and then share
their findings freely with the world.

★

For the men of the *Glacier*, Van Allen and his rockoons provided the
main excitement on a voyage that had otherwise been thoroughly routine. As Lieutenant Wilson would recall, Van Allen was at once brilliant, brave, and possessed of a certain Midwest, regular-guy charm. "He
was the kind of character," said Wilson, "that most people only read
about in *The Saturday Evening Post*." Wilson was most impressed by
Van Allen's patience, whether the professor was answering the lieutenant's endless questions about his work, or directing sailors in very practical matters. No question, or man, was unimportant. Each received Van
Allen's full respect and attention.

Although Wilson counted Van Allen as "one of a kind" he actually
represented a certain type—swashbuckling man of science—that went
back, at least, to the naturalists who sailed with European explorers to
the New World in the sixteenth century. The modern era of the scientific adventurer had dawned with the publication of Charles Darwin's
enormously popular writings about his voyages to roughly the same
region—the Galápagos and surrounding waters—where the *Glacier*
sailed on that Friday in October. By the start of the twentieth century,

the attention of scientific explorers had shifted to the North and South Poles, and the upper atmosphere, which represented the frontier for Earth-bound adventure.

One of the great polar explorers, Thomas Poulter, had been Van Allen's mentor at Iowa Wesleyan College. Chief scientist for Admiral Richard Byrd's second Antarctic expedition, Poulter had a taste for big things and was never shy about pursuing new technologies. For a time he had been famous for it. In the 1930s he designed and built a "Snow Cruiser" that was fifty-six feet long, twenty feet wide, and fifteen feet high and rode on huge custom-made tires. Both bulbous and somehow graceful in design, this tugboat on wheels contained a laboratory, a machine shop, and quarters for four. It carried a small airplane on its roof and enough food and water to last a year.

The Snow Cruiser, which resembled most the future Oscar Mayer Wienermobile, had been a sensation as it toured U.S. cities prior to its deployment in Antarctica from Boston harbor. Poulter's invention was supposed to open a new era in exploration, and help America keep tabs on the Germans, who were prowling around on the pretense of a scientific mission. When it reached its destination and lumbered off the ship, the cruiser nearly tipped over. Poulter gunned the engine and the cruiser lurched forward until its spinning wheels dug four deep ruts in the ice. No amount of effort would free the beast, and it had to be abandoned. A temporary setback became permanent as Congress refused to fund repairs. Sixty-five years later, the lifeless thing remained icebound near the Antarctic outpost of Little America.[2]

As failures go, the Snow Cruiser was spectacular. It was also a perfect illustration of the kind of science that led bookish but ambitious types like Van Allen, who forever retained a sense of awe about Poulter ("one of the most creative people I ever saw") to plunge into esoteric fields like physics. The scientific age was dawning, and Van Allen could see that the hard sciences were going to play an ever-expanding role in determining the course of history. Any doubt about this fact was erased by World War II, a war that was won as much by scientists as soldiers. In the early 1940s Van Allen had helped to invent the proximity fuse, which made antiaircraft shells explode in the vicinity of enemy airplanes. He had sailed into battle aboard the USS *Washington* to see if the technology worked. Once it was accepted by gunners and admirals, the fuse helped to hasten the end of the war. In the peace that followed, Van Allen enjoyed a reputation that caused people who controlled research budgets and icebreakers to lend him their support.

Immediately after the war, Van Allen, who was fascinated by polar auroras, devoted himself to a sweeping study of the Earth's upper atmosphere using the Aerobee, and then rockoons. His study of cosmic rays, the currents in the upper atmosphere, and geomagnetism, among other things, made Van Allen one of the world's most successful geophysicists.[3]

By the mid-1950s, Van Allen knew that to get better data during the International Geophysical Year he would have to use a much bigger rocket to put an instrument-filled satellite into orbit around the Earth. He talked about this idea with some of the nation's leading rocket designers at the U.S. Army Ballistic Missile Agency in Huntsville, Alabama, who agreed that their Redstone rocket seemed perfect for the job. With federal support, a plan was hatched to rely on the Army's launcher, topped with a second- stage bundle of smaller rockets and a third stage developed by the Jet Propulsion Laboratory operated by Caltech—the California Institute of Technology—at Pasadena. The stages would fire in sequence, each boosting the satellite higher.

The plan for an American satellite was sound, but big, government-backed science is often complicated by bureaucratic infighting. In an attempt to claim the satellite mission, the Naval Research Laboratory in Washington, D.C., proposed to use its highly reliable Viking rocket as the basis for a satellite launch and began lobbying for the idea in Washington. The subsequent squabble that arose between the services played out in tense meetings in Pentagon offices. In the middle of one argument the chief of the Army's Redstone program, General J. Bruce Medaris, was told to shut up after he delivered a rant against the airplane—"rather useful in World War II, but it was just an interim solution"—and insisted that missiles were artillery pieces and therefore the Army's domain. At another point an admiral accused the others of telling "a bunch of lies" and stomped out of the room. The admiral was later ordered to return and apologize, which he did.

By the summer of 1957, the U.S. government had gotten behind the idea of a satellite—being first in space would bring America glory—but despite Van Allen's support and the Army's concerted effort to get the job, the Redstone was not selected. Concerned that outer space be considered a peaceful realm, open to all, President Eisenhower's administration assigned the mission to a Navy rocket that was not suited for use as a weapon. Not yet operational, the Vanguard was the only sizable—multiple stages or engines—rocket being developed by any branch of the military that could not carry the weight of an atomic warhead. The larger tried-and-true Army Redstone, which was created to deliver conventional and nuclear warheads, and the Air Force Thor, then nearing completion, were passed over. The three-stage Navy Vanguard rocket derived from the Viking rocket that the Navy had already used successfully. Vanguard also appealed to Eisenhower because, unlike Redstone, it would be designed and built without employing ex-Nazis. This would make a successful satellite launch a truly all-American affair.

After the satellite program was announced, its managers began receiving helpful suggestions from citizens all over America. One writer urged the Navy to rig a button at the White House, so the president could ignite the rocket. A retired riverboat captain from Mississippi volunteered his two singing mice to ride in the satellite so their chirps could be broadcast by radio to the Earth below. Neither suggestion was accepted but the fact that they were made at all showed the range of public excitement.

But as enthusiastic as some civilians were, once the satellite project was awarded, the highest official in the Pentagon actually showed little interest in Vanguard. Secretary of Defense Charles E. Wilson thought a satellite would be a "nice scientific trick" but was generally wary of expensive research programs. He starved Vanguard for attention and made sure that projects with more direct defense applications—missiles that carried warheads for example—got priority for manpower and facilities. Others in the bureaucracy dogged the project's managers with requests for paperwork. In an attempt to choke off such requests, the Vanguard team once replied to a query for data with a six-inch-thick report they called "the confidential doorstopper."

But while the Pentagon slowed Vanguard, some American officials, including the president, still felt competitive pressure to beat the Soviets into orbit. In the spring of 1957, Eisenhower complained that the nation's image around the world was threatened by the delay, but little changed. It was mainly scientists and engineers who harbored space fantasies; ordinary citizens were far more worried about the growing threat of long-range atomic warheads.[4]

★

All of the technical and political delays that had slowed Vanguard came to Van Allen's mind on the night of October 4 when Larry Cahill rushed into the officers' quarters to tell him that Armed Forces Radio was reporting on *Sputnik*. It was around six o'clock in the evening. While Van Allen was excited by the news, a long day on deck beneath the equatorial sun had left him tired and hungry. He went to eat in the officers' dining room and looked at a movie that was being shown to the crew. Curiosity then led him to the ship's radio room, which was just below the bridge, to see if any more news had come across the ship's Teletype.

In the radio room Van Allen found David Armbrust, earphones on, hunched over a radio. Off-duty when word of *Sputnik* came, Armbrust had rushed to the radio shack and moved a sailor aside so he could try to catch the satellite's signal as it passed overhead. He was joined by Lieutenant Wilson and by a federal scientist named John Gniewk, who had hitched a ride on the *Glacier* to reach a research station in Antarctica. The four men crowded into the small space and waited in silence. Finally, at about 9:20 P.M., Armbrust heard something. It was a high-pitched beep that sounded somewhat like a cricket on a summer night.

Once they convinced themselves that they were, indeed, hearing the satellite, the men in the radio room suddenly realized that they should record the signal. Gniewk ran to his quarters to fetch a small tape recorder while Van Allen went to get an oscilloscope to generate a visual display of the wave made by the signal. Returning quickly, they managed to get their equipment going and capture about five minutes' worth of beep-beeps before the sound faded away.

When the radio signal finally disappeared, Van Allen quickly calculated that *Sputnik* would return in about ninety-five minutes. The signal came back exactly when he said it would. The only puzzle in their observations was that they then heard the beeps for much longer—more than twenty-five minutes—than Van Allen expected, given the satellite's speedy journey from horizon to horizon. He figured the radio waves were bouncing off the ionosphere like shortwave signals. As experienced radio operators know, radio waves will bounce off of charged particles in the ionosphere, and return to Earth in a repeating pattern that can carry them great distances beyond the horizon.

After the third set of satellite signals came across the radio, Van Allen dashed off a telegram for IGY officials in Washington, announcing that he had heard the satellite signal and giving David Armbrust credit for discovering it. He waited to hear the beeps one more time before going to bed. Others continued listening, and reading the news that came in over the Teletype.

In his field notebook, which he kept for posterity, Van Allen recorded what happened on the night *Sputnik* soared far above the highest point he had ever reached with a rockoon. He was impressed by the size of the satellite, writing "Wow!" next to a note about its dimensions and weight. He recognized the propaganda coup realized by the Soviets and their "Brilliant achievement!" And, unable to resist a private "I told you so," Van Allen added that Sputnik had proved that the administration in Washington had been wrong to reject the reliable Redstone, which he preferred over the new Vanguard, for America's satellite program. He believed that the Redstone rocket, which was already tested and refined, could have put the United States in space in 1956.

Although he recognized that the United States might suffer a loss of prestige, Van Allen knew that *Sputnik* posed no direct threat to American security. He wasn't so much afraid of the satellite as he was delighted by it. Van Allen expected that *Sputnik* would create a sense of crisis in Washington and that great sums of government money would be spent in response. Much of this cash was going to flow to the big men of science who, in the tradition of the Snow Cruiser, could mobilize ideas, people, and machines to accomplish great things.

As one of the leading figures in geophysics, Van Allen would be expected to join the scramble for contracts. Instead, he was in the middle of the Pacific Ocean, stuck aboard the *Glacier*, weeks away from the next port of call in New Zealand. "Causes me to be very sorry to miss the inevitable reconsideration and perhaps marked changes in the U.S. program," he wrote in his field journal. "May be a genuine loss of opportunity for us."[5]

★

Bobbing along in the middle of a dark Pacific night, alone and receiving limited information, Lieutenant Wilson and Petty Officer Armbrust had more immediate concerns. They wondered if they had been the first Americans to capture Sputnik's beep-beep as it circled the globe. Nearly fifty years later, both men would still think it was possible.

It was not. Earlier on that same night, technicians working for RCA on Long Island had picked up the *Sputnik* signal at a radio observation station near the town of Riverhead. But even these earwitnesses may have come in second to a group manning the top secret U.S. Air Force Security Service Operations Center in Darmstadt, West Germany.

Far more sensitive than anything aboard the *Glacier*, the radio receiving equipment operated by Airman Bradford Whipple in Darmstadt was built around an antenna field that was a full 150 acres. It could hear everything from Soviet gunners in tanks on maneuvers in Bulgaria to Aeroflot crews filing flight plans to Leningrad. A patriotic young man from the Boston area, Whipple felt he was performing an important Cold War duty, monitoring the Red Menace. Schooled in the Russian language and culture, he took the job so seriously that he even kept track of the cars with yellow and red Soviet license plates that he saw on the highways around Darmstadt, writing down the numbers of those that cruised near the antenna array and reporting them to U.S. authorities.

On the night of October 4, 1957, Whipple was on duty as a Russian voice intercept operator and transcriber at the secure Operations Center, filling in wherever he was needed. The work could be boring and as was his habit, a radioman whom everyone called "Hogjaw" twisted the dials to see if he could pick up the broadcast of the Yankees-Braves World Series game. (When sport broadcasts weren't available, he searched for country music.) According to Whipple, who recalled these events decades later, Hogjaw's less than rigorous devotion to his duties tended to attract extra attention from annoyed supervisors. On this night when a sergeant drifted over to check on him, Hogjaw showed him a printed recording of some strange Morse code signals—"dah-dit-dah"—that he had homed in on. This sent the sergeant scurrying to report to his superiors.

When the sergeant returned, he said that higher authorities wanted the signal monitored continuously. He yelled for Whipple to go to an operations area called the Radio/Telephone Section, find the signal, and record it on audiotape. Hogjaw continued his Morse code transcription while Whipple, who had proved himself a bit more reliable under pressure, used the most sensitive antennae in the center's array to follow the mysterious signal's voice. In minutes it disappeared, but Whipple continued to monitor the frequency. By the time it returned, about an hour and a half later, word of a possible Soviet satellite launch had reached Darmstadt.

Because Whipple and the others in his unit heard *Sputnik* before the Soviets notified the world of what they had done, it's likely that they were the first Americans to hear it. Eventually Whipple would have another, far more dramatic encounter with *Sputnik*, but that event would not occur for nearly a year. For the time being he would just track its orbits and seethe over the fact that the Soviets had beaten America to space.[6]

★

On the night of its launch, as radio stations broadcast news of its orbit, people around the world lifted their eyes to the heavens and tuned their radios in hopes of connecting with *Sputnik*. It's not likely that anyone saw the satellite, but they could have glimpsed its booster rocket, a much larger object that was also orbiting the earth. It was visible when it was in the so-called twilight zone at the horizon just after sunset or before dawn. However, because the Soviets had wisely chosen a frequency that could be received by ordinary shortwave radios, millions could hear *Sputnik*'s radio signal with just a little effort.

In Lima, Peru, reporters at one local paper got the news about the satellite via Teletype machine and immediately raced out of the city to a radio base that the U.S. Army had recently established in a mountain clearing. Set on more than twenty-three acres of flat, elevated land, the

Minitrack station was one of many that had been built as part of the Vanguard program, which was progressing slowly due to technical problems. (Others were scattered from Australia to Cuba.) At all these stations, which were equipped with sophisticated antennae and receivers, observers had spent months waiting out Vanguard delays, preparing to monitor an American satellite that had yet to fly.

On *Sputnik* night, the Americans at the Lima Minitrack outpost initially resisted their visitors and sent them away. Then they thought better of their decision to ignore the satellite news and called their boss, who was in the city. Excited by the news, Chester B. Cunningham, chief of the station, raced back and fired up the equipment, although he was not sure what he was supposed to be tracking.

At about midnight, a different group of reporters from Lima appeared and confirmed and expanded on the story told by the first. After much back-and-forth (some of the Peruvians returned to their office to retrieve more information), Cunningham realized what they were saying about the frequencies used for the *Sputnik* signal. He donned headphones and began working the dials on a radio. What happened next was reported in a memo he sent to his Project Vanguard bosses:

> *At 1:58 AM on October 5 we got the characteristic signal. However I did not believe it to be a signal from the satellite. In 10 minutes it faded out. I then predicted to the reporters that if the signal were from the satellite, it would come back at 2:50. Almost precisely to the minute the signal reappeared. In five minutes the reporters took off for Lima like bullets. We stayed around until 4 AM and got a third passage, when we went back to Lima to get some sleep . . . we got back to the station at 8 AM on Saturday, and from then on it was bedlam.*

Over the weekend big crowds gathered at the Minitrack station outside Lima to inquire about *Sputnik*. A tape recorder was brought from the U.S. embassy and Cunningham used it to play the beep-beep for the Peruvians at the gate. Cunningham also posted notices informing visitors about the satellite, and used the chore of watching *Sputnik* as a training exercise for his men.

Through it all, Cunningham felt uncomfortably humbled by the blow *Sputnik* delivered to America's standing in the world. In his ultimate report on these activities, right before he requested eighteen hours of overtime pay for the event, Cunningham noted that the Peruvians actually expressed "sympathy" for their giant neighbor to the north. Cunningham said he was worried that America may "have lost some face."[7]

★

A little loss of face didn't seem to bother the Eisenhower administration. At the highest levels in the American government, the first response to the Soviets' achievement was generally calm. Many had been prepared for the news. In fact, just hours before the rocket blasted off from Baykonur, a senior Central Intelligence Agency official had actually forecast the Soviet satellite success. In a long-planned briefing, Herbert Scoville Jr., deputy director of the CIA, told a White House panel on science, "It wouldn't surprise us if such an announcement [of a satellite success] came at any time."

When the prediction became a reality, President Eisenhower may have been annoyed that his Vanguard had been beaten into orbit, but he was hardly concerned from a military standpoint. "We knew that *Sputnik* was not a danger," recalled physicist Herbert York, who was one of Ike's science and technology advisers at the time. "And we knew we could catch up." Indeed, York believed that while the Soviets deserved congratulations, Americans had no reason to panic.

This attitude of equanimity was shared by American IGY officials who were gathered on *Sputnik* night at the Soviet embassy in Washington for a party. After taking a phone call from his bureau, *New York Times* reporter Walter Sullivan bounded up the embassy stairs to the ballroom to deliver the news. When it was announced to the crowd, the Americans warmly applauded their hosts.

The next day, at a meeting in Washington of the National Academy of Sciences, a Soviet space expert named Anatoli Blagonravov happily agreed to a change in the scheduled program to brief the crowd on the satellite. As he puffed a pipe, the white-haired Blagonravov made a drawing of *Sputnik* on a blackboard—it looked like a flying basketball trailing four antennae—and reassured his colleagues that it was nothing more menacing than a test vehicle launched to prove that the R-7 could put an object in orbit.

Although they may have envied the USSR's success, the Americans who listened to Blagonravov offered mainly admiration. When the Soviet scientist asked if the Americans might provide his delegation with a radio so they could hear the satellite's signal, Vanguard director John P. Hagen supplied something even better. He brought a tape recorder onto the stage and switched it on to let the beep-beeps captured by American monitors fill the hall.

"That is its voice!" blurted Blagonravov.[8]

★

If the voice of Sputnik thrilled the scientists in Washington, it thoroughly annoyed certain ambitious and competitive men in Huntsville, Alabama, where the Army Ballistic Missile Agency operated on the grounds of the sprawling Redstone Arsenal. In this once sleepy cotton-trading center, a band of former enemies—U.S. military officials and ex-Nazi rocketeers—had been working on missiles to match and surpass the Soviet's best. They were directed by two leaders—one a U.S. Army general, the other a former SS officer—who tended to fight against the limits imposed on them by bureaucracies, ideologies, or even orders from Washington.

Poetically enough, one of these men, General J. Bruce Medaris, had almost been killed by a rocket designed by the other, former Nazi rocket chief Wernher von Braun. The incident had occurred near a battlefield in France fourteen years earlier. While working in a mobile office, Medaris heard such a loud explosion he thought that an ammunition dump had blown up. Racing outside he found dazed men wandering near a big smoking hole in the ground. Down in this pit he recovered tubing and valves from one of von Braun's rockets, which had apparently malfunctioned on its way to Great Britain.

Years later Medaris would tell this story with the cavalier tone of a man who wanted you to know he had been toughened by combat, though in fact he had seen very little real fighting. Originally trained as a horse soldier (but too late to fight in World War I), Medaris had spent some time in civilian life. But after returning to the service for the Second World War he had made the Army his career as an ordnance officer. Successive promotions had landed him at Redstone, where he commanded the development of the Army's next generation of ordnance—rockets equipped with warheads.

By the fall of 1957, gray was invading the fifty-five-year-old Medaris's bushy black mustache and slicked-back hair. Surgery for prostate cancer the year before had left him a bit slumped and potbellied. But the general still presented himself with a firm, military attitude, a style consistent with the old-fashioned swagger sticks he collected and carried on a regular basis. Having joined the military when horses still went into battle, Medaris's style owed as much to the nineteenth century as it did to the twentieth. He respected rank and was so egotistical he once told a national magazine, "No human being, without the guidance of the Lord, could have been right as much as I was."

As he approached the end of his military career, the general had become devoted to keeping the U.S. Army relevant in an era of changing technology. Airplanes and missiles were rising in importance, and he worried about the Air Force and Navy, which wielded both, getting too far ahead. In Medaris's imagination, one day the Army could be equipped with one kind of rocket that could catapult small teams of men into enemy territory, and others that would serve as ultra-long-range artillery. This view was consistent with a little publicized variant of American strategy that envisioned the limited use of nuclear weapons in war. A chief proponent of this idea was Medaris's friend at the Pentagon Lieutenant General James "Slim Jim" Gavin, who saw nuclear weapons as an antidote to any enemy's superior manpower or conventional arms. Chief of Army Research and Development, Gavin had led the 82nd Airborne during World War II and had been the youngest general in the Army since Custer.[9]

Medaris's leadership partner at Redstone, von Braun, was movie star handsome with a chiseled chin, thick greased-back hair, and piercing blue eyes. Von Braun had become besotted with the idea of space travel as a boy growing up in Prussia. By the time Hitler annexed Austria in 1938, von Braun was, at age twenty-six, one of Germany's leading rocket experts. He was also, in the eyes of his then colleague Willy Ley, a fair-haired, six-foot-one-inch "perfect example of the type labeled 'Aryan Nordic.'" Soon he would become director of the Reich's military missile program, charged with developing weapons that could reach Great Britain from bases on the European continent. (Ley, a writer and early member of Germany's national amateur rocketry society, fled the Nazis for America in 1935.)

Always politically keen, von Braun learned how to satisfy a government bureaucracy while also indulging his dreams of space. He joined the Nazi party and its notorious SS paramilitary and pushed hard to fulfill Hitler's demand for rockets to be fired at civilians in London. His effort would culminate in the V-2, the world's first guided ballistic missile. What he learned about the art and science of space travel came as a side benefit to this military work. He tried to hide the fact that space was his personal priority. And according to one oft-repeated story from this time, this obsession almost got him killed.

The way the tale has been told, von Braun was traveling on a German train when he fell into a conversation with a woman who happened to share a semiprivate compartment. He talked about his dreams of space and how his true passion was exploration, not explosions. The woman was an SS spy. Soon Gestapo agents arrived at von Braun's home in the middle of the night. Certain that the rocket chief was diverting his time and energy into his real obsession, research on space travel, they arrested him on charges that he was undermining the war effort. Equivalent to treason, the accusations could have led to the death penalty, but two of von Braun's allies in the army—armaments chief Albert Speer and missile leader Walter Dornberger—intervened with Hitler himself. After two weeks von Braun was freed on the führer's order.

This account of von Braun's arrest would stand for years until it was refuted by American historian Michael Neufeld. The facts, Neufeld reported, show that von Braun had been caught in a contest between the SS and the army over the future of the missile program. He was arrested after he refused to cooperate with SS chief Heinrich Himmler's takeover attempt. Neufeld did confirm, however, that von Braun's release came about as a result of Speer's intervention. And obviously the young engineer must have been terrified by the experience.

A few months after he was arrested and freed, one of von Braun's V-2 rockets, deployed with a mobile launching group, was finally fired from Belgium against England. It struck a London suburb, killing three civilians, and began a season of missile attacks that would last until March 1945. Roughly 1,500 of these rockets, each carrying approximately 2,000 pounds of explosives, were used against England, and they killed more than 2,500 people.

Britons would regard the rockets as indiscriminate terror weapons, which they were. The men of the German rocket team, who considered their attacks a justified response to Allied bombing raids that killed thousands of civilians, celebrated their first successful firings with champagne. Justified or not, the rocket attacks were not militarily decisive. By the time they began, the Allies had already liberated Paris and the outcome of the war was certain. When surrender was imminent, von Braun thought hard about where he might find a way to continue his work. As one of his colleagues would later explain, "We despise the French; we are mortally afraid of the Soviets; we don't believe the British can afford us; so that leaves the Americans."

After deciding that the rich and amiable Americans might take him in, von Braun loaded a caravan with blueprints, equipment, and hundreds of colleagues and set out to find them. The Americans who took the Germans into custody knew that their prized scientists were tainted by the evil of the Nazi regime. The V-2s had been constructed by thousands of slave laborers—many from Nazi concentration camps—in conditions that recalled Dante's Inferno. Thousands of V-2 slaves—more than the missiles killed when they were fired—died of starvation and disease or at an executioner's hand.

Decades later, proof that von Braun had personally guided the selection of concentration camp prisoners for his missile shops would be made public. But at the time he was taken into custody, the U.S. Army wanted him and the other German rocketeers to build missiles for the United States. Despite opposition from others in the government, the military whitewashed the rocket men's past and they were treated as prized possessions rather than war criminals. The War Department's public notice of this project, issued October 1, 1945, said the move would be temporary, however, and only those scientists and technicians deemed "outstanding" would be included.

Brought first to Fort Bliss, near El Paso, Texas, von Braun and his V-2 team became vital to the Army's rocket development program. Initially confined like prisoners of war, the Germans were quickly granted their freedom, and authorities allowed them to bring their families to the United States. In 1947 von Braun went back to Bavaria to marry his eighteen-year-old cousin Maria von Quistorp, who would become a kind of first lady in the small society that the Germans were building in America. By 1950, when the group was moved to Alabama, the notion that they were only temporary visitors was long forgotten. In five years von Braun would become a U.S. citizen and, in the great American tradition, was well on the way to reinventing himself.[10]

Through public appearances, his own writing, and press interviews, von Braun washed himself clean of his Nazi past. Book publishers played their part, including von Braun in three popular anthologies about space. But the editors of *Collier's* magazine were especially helpful in this task, offering their 3 million subscribers articles about space by von Braun himself and devoting extensive coverage to his enthralling ideas about the future. These pieces, illustrated with fantastic drawings of planets, spacecraft, and astronauts, made it seem as if science fiction fantasies were on the verge of becoming reality. They led to more press interviews and appearances on television.

By 1955, von Braun was such a popular figure that he was featured prominently, along with others from his Nazi days, in a Walt Disney TV series about space exploration. (Two installments aired in 1955. A third would be televised in late 1957.) The series, which featured some rather psychedelic animation—rock-eating animals and walking space plants— was directed by Ward Kimball, creator of Jiminy Cricket and fresh off the job as directing animator of *Peter Pan*. Kimball and Disney also relied on von Braun's advice as they designed the famous Disneyland theme park in Southern California. When it opened in the summer of 1955, much of the section called Tomorrowland, including its Moonliner attraction, owed its sense of style to the V-2 and other von Braun conceptions.[11]

Disney's embrace, and all the other publicity, contributed to von Braun's status and confidence, and encouraged his dream of human spaceflight. By 1957, he and General Medaris dominated the rocket center at Redstone Arsenal, which in turn played a big role in the development of northern Alabama. As the Army Ballistic Missile Agency had evolved at Redstone, it created hundreds and then thousands of new jobs and a more diversified and wealthy local economy. But the newcomers were not welcomed by all. Some locals shunned the Germans, who for comfort, built their homes together in a neighborhood that locals called the German Colony. But by the mid-1950s people in Huntsville generally took pride in their community's status as the self-proclaimed "Rocket City" and embraced the region's role on the front lines of the Cold War.

The Army group had raced to develop longer-range missiles that could serve the dual purposes of nuclear attack and satellite vehicle. Only orders from Washington had prevented them from putting an object in orbit in 1956. When the Eisenhower administration then chose the Navy team to launch the first American satellite, restricting the Army to shorter-range weapons development, the Alabama rocketeers appeared to comply. But in fact they were so certain Ike was wrong that they secretly continued satellite work and they quietly reserved one of their babies, Missile 29, just in case they were called to launch on short notice.

On the day *Sputnik* was launched, Medaris and von Braun hosted visitors from Washington, including the new secretary of defense, Neil H. McElroy, and the Army's chief of research and development, James Gavin. McElroy had come to government service from Procter & Gamble, where he had worked for more than thirty years. Gavin, former commander of the 82nd Airborne in World War II, had solidified his status as a folk hero by dating Ernest Hemingway's famous journalist ex-wife, Martha Gellhorn. He would become known as a sharp critic of Eisenhower's reliance on nuclear deterrence and cost-saving cuts in troop strength.

Over lunch that day at the Redstone Arsenal, Gavin and von Braun had discussed the possibility that the USSR might put a satellite into orbit. "Based upon all available intelligence, the Soviet launching was imminent," Gavin wrote months later. They agreed that a proper response would include the development of some kind of "satellite interceptor" along with a "militarily useful satellite" for the American side. This plan fit Gavin's view of space as a military realm occupied by manned stations, from which America might deploy or manage missiles, drones, and surveillance craft to win an "Earth war" against an enemy such as the Soviet Union.

That night, Gavin joined von Braun, Medaris, and the new secretary of defense at the Redstone Arsenal officers club. The hum of the crowded social gathering was broken when it was announced that the Soviets had put *Sputnik* into orbit. Though he fully expected it, the normally unshakable Gavin felt sickened by the news. He believed that his greatest fear, that the Soviets were pulling ahead militarily, was coming to pass. "I felt crushed," he would recall, "and wanted to take a long walk to think the situation over alone."

Unlike Gavin, who felt discouraged, Bruce Medaris furiously blurted, "Those damn bastards!" He could have been talking about the Soviets, the Navy team that had fallen behind on Vanguard, or the higher-ups in Washington who had prevented his team from going to space. Certainly one of *those* bastards was Defense Secretary McElroy's immediate predecessor, Charles Wilson, who had been in charge when the Navy rocket was picked for the first American satellite.

While Medaris vented his emotions, von Braun bore down on McElroy. This was one of those rare moments when events seemed to justify a dramatic response and von Braun demanded that "for God's sake" his team be turned loose on the satellite project. Predicting that McElroy would find "all hell has broken loose" when he returned to Washington, von Braun pressed him to keep one thing in mind. "We can fire a satellite into orbit sixty days from the moment you give us the green light."

The next morning, the mood at Redstone was so gloomy it seemed like the whole place had been drenched in defeat. Before he departed, McElroy heard much more about the Army's reliable rockets from a team of younger officers and engineers. With the secretary scheduled to depart at noon, the Redstone group gave McElroy a quick tour of their facilities, making sure he saw they already had the hardware on hand to go into space. They didn't get immediate permission to restart their satellite project, and when McElroy got back to the capital he told reporters only that he was going to resolve the "bottlenecks" that had kept America earthbound. However, General Medaris, ever confident, told von Braun he could assign some workers to prepare Missile 29. If they never actually got the green light, he said, the bookkeeping could be adjusted so the expense of this jump start would never be noticed.[12]

TWO

New Moon Worries

Oh little Sputnik flying high
With made-in-Moscow Beep
You tell the world it's a Commie Sky
And Uncle Sam's asleep

You say on fairway and on rough
The Kremlin knows it all,
We hope our golfer knows enough
To keep us on the ball

MICHIGAN GOVERNOR
C. MENNEN WILLIAMS

In time, nostalgic impulses and the soothing nature of memory would create a picture of America in the 1950s as a place where everyone agreed on almost everything. Politics of all kinds—social, electoral, even sexual—were practiced in a relentlessly agreeable way. Everyone lived in a stable heterosexual family that valued cleanliness and digestive regularity. Everyone belonged to a supportive community. Everyone stood with their amiable old, golf-addicted president, Dwight D. Eisenhower, the bald grandfather who had defeated Hitler.

But of course, America in the 1950s was not the monochromatic dream so many would one day imagine. To believe this, you would have to ignore the nation's fascination with Alfred Kinsey's frank reports on the varied sexual habits of men and women, the growing influence of so-called Beat writers like Jack Kerouac, and the rise of rock 'n' roll, which was blamed for, among other things, the perceived scourge of juvenile delinquency. By 1957, Elvis was already Elvis, scandalizing the old and thrilling the young.

In the supposedly quiet 1950s, a steady rise in divorce brought upheaval to the nuclear family while the civil rights movement began a national fight over the inequalities in the larger society. C. Wright Mills laid out the basic anti-establishment critique of the 1960s in a book called *The Power Elite*. At the same time, the abuses of Republican senator Joseph McCarthy and his chief aide, Roy Cohn, proved that politics in this well-mannered decade was as nasty as ever.[1]

Both a product and beneficiary of Cold War fear, McCarthy had persecuted the administration of Eisenhower's predecessor, Democrat Harry Truman, with accusations about communists in the government and negligence in the pursuit of the Cold War. (Oddly enough, long before the satellite flew he called some of those he baited "sputniks" because he considered them communist "fellow travelers.") Almost all of McCarthy's charges were baseless and in his investigations he found precious few real communists. But before he was stopped, his assault, conducted through committee hearings, speeches, and leaks to the press, helped the GOP win the White House for Ike and gain more than a few seats in Congress.[2]

The political damage done by McCarthyism was felt most keenly by Democrats, while Republicans emerged from the witch hunt era with a tough-on-Reds image. This helps to explain why, as reporters and editors prepared extensive reports on the *Sputnik* "crisis" for the thick Sunday editions of the nation's newspapers, the president stuck to his regular weekend schedule. He relaxed at his gentleman's farm in Gettysburg, played golf with his neighbor, and visited with his grandchildren.

Ike left it to the men in his administration and allies in Congress to handle the nation's new moon worries. (The press had immediately labeled *Sputnik* a "moon.") Rear Admiral Rawson Bennett, who was involved in the Vanguard program, stepped forward to describe *Sputnik* as a "hunk of iron almost anybody could launch," and Republican senator Alexander Wiley of Wisconsin insisted it was "nothing to worry us." White House press aide James Hagerty told reporters that Sputnik was of "great scientific interest" but refused to grant that it had any military significance.

With all he knew about military matters, Eisenhower truly saw no reason for alarm. At the start of his presidency he had focused intently on the threat of a Soviet nuclear attack and asked both scientists and intelligence experts to separate the facts about the USSR's capabilities and intentions from the propaganda. For years the Soviets had done everything possible to suggest they possessed more powerful forces than were actually in place, down to having bomber squadrons fly over Moscow in circular routes to suggest a much bigger force.

Frustrated by the lack of accurate information, Eisenhower had approved a secret project, urged by the president of the Massachusetts Institute of Technology, James R. Killian, to develop a spy plane that could pass over Soviet territory at such a high altitude it would be safe from attack. The plane that was built, at Lockheed Corporation's secret factory nicknamed the Skunk Works, was a super-light, glider-like craft that could fly at 70,000 feet for nearly 5,000 miles without refueling. The magic that allowed for this plane to be built could be found in new light materials and a design that allowed for a wingspan — 102 feet — nearly twice the plane's length. To hide any suggestion that the plane was special it was dubbed the U (for utility) 2. There was no U-1.

The cameras built to travel aboard the U-2 were similarly advanced. They produced amazingly detailed photos of the ground from extraordinary altitudes. Eisenhower had first seen this when the chief of the program showed him photos of the Gettysburg farm, taken from 60,000 feet, which showed each one of the president's cows. Once the spy planes were deployed over the USSR they proved that a suspected Soviet advantage in bombers — called the "bomber gap" — was a fiction. (Eisenhower, who had already begun calling the hype about Soviet strength "the numbers racket," knew the American fleet of more than a thousand B-47s and B-52s was vastly superior.) The postcards from the edge of the atmosphere also suggested that other claims of Soviet military superiority were similarly inflated. Of course none of the U-2 information could be shared with the public, but it certainly helped the president put *Sputnik* into proper perspective.

The U-2 performed in secret a job that Eisenhower had previously suggested be done out in the open, by both sides, at far less risk and expense. His 1955 Open Skies proposal had called for regular surveillance flights that would have given the Soviets full view of America's military activities and provided the same for the United States in the USSR. This simple scheme, which could have calmed Cold War tensions and provided reassurance to both sides, had been rejected by Nikita Khrushchev. But in putting the plan forward Eisenhower emphasized the vital importance of accurate information and won himself points as a peacemaker.[3]

★

On the night the Soviet satellite was launched, U.S. senator and political dramatist Lyndon Baines Johnson finished dinner and took a stroll outside his Texas ranch with some guests. One of these, Senate lawyer Gerald Siegel, would recall that they set off for the nearby home of Johnson's "Cousin Oriole," who often hosted parties for the senator. On the walk, Johnson decided he would lead an investigation of missile and space issues and that he would immediately contact other Senate leaders for support.

In a memoir published in 1972, Johnson, whose basset hound face gave him a weary-wise presence, would paint a romantic, lonesome cowboy kind of scene as he described the night that Sputnik first circled the Earth:

> As we stood on the lonely country road that runs between our house and the Pedernales River, I felt uneasy and apprehensive. In the open West, you learn to live closely with the sky. It is part of your life. But now, somehow, in a new way, the sky seemed almost alien. I also remember the profound shock of realizing that it might be possible for another nation to achieve technical superiority over this great country of ours.

No one could say for certain what was in the senator's heart on that starry night. The poetry in LBJ's memoir came after many years and was aided by an editorial team that included the gifted writer Doris Kearns, who would later win a Pulitzer Prize for biography. However, Johnson's actions left no doubt that, as usual, both policy and politics were on his mind as *Sputnik* whirled overhead. Indeed, as soon as he was back inside his house he began working the phone to prepare for a high-profile investigation of America's defeat in space.

By Sunday night, the American people (or at least those who paid attention to affairs in Washington) would understand that Johnson and the Democratic leaders of the U.S. Senate intended to find out why the Russians got into orbit first. Early Monday morning Johnson was on the phone with Eileen Galloway of the Congressional Research Service, who was an authority on both defense and government organization. Galloway had recently completed a paper on guided missiles in foreign countries. "Eileen," said Johnson, "I want to make me a record in outer space and I want you to help me." Galloway soon began racing to prepare for Senate hearings on space, satellites, and America's response to the Soviet challenge, which Johnson would conduct. "If you were working for Lyndon Johnson," she would recall when it was over, "everything had to be done in a hurry."[4]

Although Johnson moved to put himself at the center of space politics, his comments were, at first, moderate. Others were less deliberate in their response to the satellite news. In Kansas City, LBJ's fellow Democratic senator W. Stuart Symington issued a statement blaming the Eisenhower administration for allowing the Soviets to gain "superiority in the all-important missile field." (A former secretary of the air force, Symington was a LeMay-style hawk devoted to Pax Atomica.) Senator Henry Jackson of Washington called *Sputnik* "a devastating blow to the prestige of the United States." Jackson also accused the president of hiding the fact that he had starved America's satellite program for funds. (Eisenhower had cut defense spending by about 20 percent in order to produce a federal budget surplus.)

In the rush of early reactions, even one of Ike's loyal Republicans turned negative. The difference was that Styles Bridges of New Hampshire preferred to scold the American people, rather than the president, with an outburst worthy of Cotton Mather. "The time has clearly come," preached Bridges, "to be less concerned with the depth of the pile on the new broadloom rug or the height of the tail fin on the new car and to be more prepared to shed blood, sweat and tears if this country and the free world are to survive."

Bridges's critique would be seconded at home and around the world by those who believed they saw great flaws in the American character, which were responsible for its defeat in the competition for the first satellite. Ironically enough, this moralizing was expressed in similar ways by God-fearing preachers at home and godless communists abroad. At a conference held in Spain just days after the Sputnik launch, Soviet satellite expert Leonid Sedov lectured his American counterparts on the shortcomings of their home country.

> *America is very beautiful, very impressive. The living standard is remarkably high. But it is very obvious that the Average American cares only for his car, his home, and his refrigerator. He has no sense at all for his nation. In fact, there exists no nation for him . . . he also has no sense for great ideas which take as long as a number of years to achieve, and which do not pay off immediately. He just does not feel attracted to them. Russians do!*

If the moralists and communists had been alone in their criticisms, the Eisenhower administration's plea for calm might have worked. Instead, the press found plenty of respected scientists and other seemingly wise men (this was a time when few women were granted such status) to join the chorus of complaint. On Friday night—*Sputnik* night—several prominent experts told *The New York Times* that the administration had blundered when it picked the Navy over the Army for the U.S. satellite project and then failed to provide the team with enough funds. Perhaps because they still wanted to receive government money for their own projects, most of these critics spoke anonymously. But one, Dr. Israel Levitt, Ph.D. of the Fels Planetarium in Philadelphia, went on the record as he blamed the government's "incredible stupidity" for letting the Soviets lead the world into space.

Along with the criticism from scientists came warnings from other experts who were sure that *Sputnik* was sending out secret messages. Observers at Caltech and the Naval Research Laboratory each made this claim within twenty-four hours of *Sputnik*'s launch. The men scanning the airwaves for RCA on Long Island said they believed they heard a secondary tone underneath the original *Sputnik* signal. An amateur radio enthusiast in Massachusetts reported that he heard signals that sounded like the dots and dashes of Morse code. From Woomera in the

Australian Outback came a report that the Soviets had beamed a message to the satellite and that *Sputnik* had replied. John Dowling, who was in charge of the American IGY tracking station at Woomera, said he couldn't explain the messages because "we haven't got the code."

The first visual observation of the satellite would be claimed by Gordon Little, Ph.D., at the University of Alaska, who viewed Sputnik for about five minutes as it passed overhead and, in a comment that should have delighted the Chief Designer, noted that it was so "surprisingly bright" he could see it without a telescope or even binoculars.[5]

Little's observation would be confirmed by the Smithsonian Astrophysical Laboratory outside Boston, which quickly became the main source for reporters seeking the most authoritative scientific views on *Sputnik*. Soon the director of the observatory, Fred Lawrence Whipple, and his colleague J. Allen Hynek were offering formal briefings to groups of journalists who received a crash course in astrophysics and astronomy.

With receding hair, a beard, and a pipe frequently clenched between his teeth, the forty-seven-year-old Hynek fit the tweedy scientist stereotype reporters expected and he instantly became a favorite source. Having established a network of amateur stargazers as part of the ICY, Hynek maintained contact with hundreds of reliable spotters, each equipped with a telescope. When the *Sputnik* launch was announced he had alerted this group, suddenly called "the moonwatchers," and they had helped him to confirm Gordon Little's sighting.

(For some of Hyenk's moonwatchers, who were mostly males, satellite tracking offered a chance for father-son bonding. In Paterson, New Jersey, Philip DelVecchio and his fifteen-year-old son, Robert, spent hours together tuning in to Sputnik's radio beacon and scanning the sky with a telescope wedded to a camera. They sent photos and recordings of the radio signal to U.S. officials and made reports to observatories in America, Switzerland, and Ireland.)

In the days after the *Sputnik* launch, Hynek would become one of the main voices of space science, acting like an ancient priest divining some celestial event. He had performed this duty before. In 1951 he alerted the nation to a sunspot so enormous it could be seen without a telescope by anyone peering through unexposed photo film. Three years later, in June 1954, he organized observers for a rare solar eclipse. A month after that, a team he led explained to the world what makes starlight twinkle. Unstable elements of the atmosphere are to blame.

Hynek was a space enthusiast, and he used this historic event and his access to the press to fan public interest in rockets, missiles, satellites, and so forth. Absent from his comments, however, was any reference to his research on flying saucers—UFOs, Unidentified Flying Objects. Unbeknownst to most Americans, Hynek was at the time a leading consultant for Project Blue Book, the Air Force program established to analyze reports of mysterious craft in the sky. Blue Book had been started in 1952, at a time when flying saucers were reported by sources across the country, including pilots who insisted they saw some in the skies over the nation's capital. Blue Book was ongoing in 1957, and its staff was far from ready to issue a public report.

Altogether, the news that appeared in the press immediately after the launch of *Sputnik* was exciting, disturbing, and intriguing. All the major papers in America gave the event extensive coverage and broadcasters on the networks filled the airwaves with reports. Amateur radio pranksters got into the action quickly, sending out fake signals. One even accessed the same frequency used by *Sputnik* to announce, in plain English, "This is the satellite."[6]

As so often happens when the government seems surprised by events (although the administration wasn't truly caught off-guard), political commentators immediately declared that the intelligence agencies had failed. The reasons most widely cited for this failure were psychological. Americans had been deluded by the stereotypes they held about Soviet backwardness, the superior wealth of the United States, and the shortcomings of a communist-style command economy. While Soviet-style central control didn't foster creativity, it certainly could accomplish big things.

The political and scientific noise made 'round the world about *Sputnik* was so loud that the Chief Designer heard about it from the pilot of the propeller-driven plane that took him from Tyuratam to Moscow the night of October 5. Leaving the controls, the pilot walked through the cabin to find the sleep-deprived hero and tell him "the whole world's abuzz."[7]

★

Pravda may have been slow on the pickup, but once Khrushchev understood that he had the world's attention and had shaken the United States, he felt like rubbing it in. Fourteen months earlier he had been on the other side, as U-2 spy planes violated his country's airspace, flying just out of reach of Soviet fighter jets and their air-to-air missiles. To add to the insult, Khrushchev had been visiting the U.S. embassy in Moscow to pay his respects on the Fourth of July when Soviet radar picked up the first U-2 flight. When he heard about the intrusion, he imagined his enemies in Washington laughing with delight. Now that his country had a craft that flew higher than the American spy planes, and could not be stopped, the chairman was eager to prove the success was no fluke.

According to a tale handed down from one of the Chief Designer's friends, when the two men talked about *Sputnik*, Khrushchev first confessed, "We never thought that you would launch a sputnik before the Americans. But you did it," he added. "Now, please launch something new in space for the anniversary of our revolution."

The anniversary was always a big deal in Moscow, but this one would mark the fourth decade of Red rule and Khrushchev intended to invite the planet's other great communist dictator, Mao Tse-tung, for the November 7 festivities. There would be the usual hours-long parade of military hardware along a route that went through Red Square. And if it could be arranged, Khrushchev hoped, there would also be a small parade routed through space over America.

The Chief Designer didn't labor over Khrushchev's request. He immediately agreed to make another satellite launch before November 7 and added an extra twist: *Sputnik II* would carry a living, breathing payload. A mutt from a kennel maintained near Moscow would be sealed in a capsule and hurled into orbit. Success would show that the USSR was on its way toward putting men in space.

Pulling off such a feat in a matter of weeks would be difficult, but the Chief Designer was energized by the challenge. And fortunately for him, the USSR had lots of dog-launching experience. Soviet scientists had been sending canines aloft since 1951, when they put a dog atop a rocket based on captured German V-2s. In time they developed a sturdy vehicle for these launches. Equipment inside recorded heartbeat, breathing, and other vital signs as the dogs rocketed up as high as 130 miles and then fell to Earth. Many actually survived. These experiments yielded valuable information for rocket teams contemplating future manned spaceflights and for those working on new jet aircraft that subjected pilots to unprecedented extremes in acceleration and altitude.

The Soviets weren't alone in this field. Americans had started slinging animals into the air in the summer of 1948 when monkeys named Albert I and Albert II gave their lives to science aboard captured German V-2 rockets launched from White Sands, New Mexico. They were followed by a mouse that was sent aloft and photographed floating weightlessly inside a tiny capsule. Then came Albert III, another monkey, who died when the parachute on his capsule failed to open as it descended.

These animal-in-rocket experiments provided valuable information, but scientists had other ways to gather data on the effects of rapid acceleration, sudden deceleration, and other aspects of space flight. At research centers scattered around the country animals as big as bears were strapped into seats and propelled down test tracks at speeds approaching 1,000 miles per hour. Rodents were floated in balloons that reached 57,000 feet. But just as with rocket flights, danger loomed in these tests too. During a windblast test conducted by the Navy at a place called China Lake in the Mojave Desert, one ill-fated guinea pig was blown through an inch-wide hole in his container, never to be seen again.

When animals were recovered and examined—often posthumously—experimenters were able to add to the small but growing base of knowledge about the effects of extreme flight conditions. A chimpanzee hurtling backward at high speed suffered brain damage and head bruises. A monkey tumbling through the air in a simulated, out-of-control ejection from a jet died, not from impact with the ground, but from the damage done to vital organs as he was jerked around in the fall.

Fortunately for the scientists, who might have faced the wrath of anticruelty societies, these animal acts had been performed without much public notice. However, human experiments were widely publicized, and both the press and the public tended to respond with approval and wonder. Some of the most fantastic feats were achieved in California at the Human Decelerator, where men suffered broken bones and black eyes riding a rocket-powered sled along a 2,000-foot-long railway. Each test run was rigged to simulate a jet airplane emergency, including crashing, spinning, tumbling, and rapid deceleration. Although pain, fractures, and other injuries were common, occasionally sled-riders reported unexpected effects such as giddiness and even boredom.

The most famous of the human test subjects was Dr. John Stapp, then a U.S. Air Force captain, who volunteered to prove that a human being could survive forty-five times the force of gravity while riding a rocket sled at Holloman Air Force Base in New Mexico. Stapp, who likely got a taste for adventure as the child of missionaries in Brazil, also flew 570 miles per hour in an open-cockpit fighter jet to prove a pilot could survive the windblast if the canopy of his aircraft blew off. Of course, he tried this feat only after some animal went first.

By the mid-1950s, Stapp's feats, conducted mainly on the seven-mile track at Holloman, had made the forty-something bachelor officer a minor celebrity. A widely published photo series of the man called "The Human Bullet" showed how his face was distorted by gravitational forces. The pictures made Stapp look like a grotesque daredevil and, consequently, helped turn him into a role model for lots of young boys.

Stapp was joined on the pedestal by other famous risk-takers. One month before *Sputnik*, an Air Force doctor named David Simons had made his way to the bottom of an old iron ore pit mine in Minnesota, where a huge balloon was being filled with 3 million cubic feet of helium. On the mine floor, which was 420 feet deep, Simons was sealed inside the capsule, which was covered in shiny foil to reflect warming sunlight, shortly before the balloon was released.

More than 130 feet in diameter, the balloon looked like an enormous white onion as it rose in the Minnesota sky. In the eyes of those on the ground it grew smaller and smaller until it looked no larger than a silvery BB, with the sun glinting off its surface.

Dangling beneath the onion, inside the capsule, Simons wore a pressurized suit. (At altitudes higher than 60,000 feet, un-pressurized blood boils.) He breathed oxygen from tanks as he passed the 100,000-

foot mark, and peered out a window at a view that no one had ever seen before. It was a magnificent sight for Simons, who suddenly realized he had forgotten to eat in the twelve hours before launch. He opened some Air Force rations—a hamburger and candy bar—and dined while gazing at his home planet.

During his thirty-plus hours aloft, Simons became the first person to see both a sunrise and a sunset across the curvature of the Earth. At night, cool temperatures contracted the gas in the balloon and it sank to 70,000 feet. As luck would have it, the descent brought the capsule close to a thunderstorm and dangerous turbulence. To escape this threat, Simons jettisoned some heavy batteries that he didn't need for the mission and, with his capsule lightened, gained enough altitude to find safety.

As the mission came to a close and Simons began to descend, monitors on the ground noted his speech was slurred and his responses to their questions were slow and hesitant. The air system in the capsule was not working well and he was being affected by excess carbon dioxide. Despite his altered condition Simon managed to land safely in an alfalfa field in South Dakota, where he quickly recovered. The only noticeable physical effect of his flight was a seventeen-pound weight loss, presumably due to dehydration.

As he reflected on his experience, Simons was filled with a sense of awe about the "fabulous view and precious opportunity" he had enjoyed inside the cramped metal box. In the days that followed this feat, his lyrical descriptions of the colors in space and stars that did not twinkle (no atmosphere to wrinkle their light) made him temporarily famous. The United States Junior Chamber of Commerce would name Simons one of the year's "ten outstanding men."

No publicity accompanied Simons's subsequent scientific papers on the flight, but they revealed a more complicated and complete picture of his experimental journey. In addition to wonder and awe, Simons had experienced severe fatigue and a decline in mental acuity. Questions radioed from the ground made him feel irritated and he was reluctant to answer. He had trouble judging his own physical condition, and consistently reported that he was in much better shape than he was, according to the instruments tracking his heart rate, temperature, and breathing.

One performance test, which called for him to make voluntary observations on the flight, showed a steady deterioration in his reports as he became less attentive and engaged. By the time Simons viewed his second sunset, he showed far less enthusiasm for the event.

Afterward, Simons would blame lack of proper sleep and the isolation of the capsule—in other words, loneliness—for his declining performance. Whatever the cause, the problem suggested trouble for future space travelers, who could be expected to overestimate their own abilities and underestimate the negative effects of long solo flights. In studying his own experience and those of others who undertook similar missions, Simons also found a pattern of self-deception—pilots thought they performed better as time passed—which he reported with a scientist's candor:

> *All operators showed intermittent periods of seriously deteriorated vigilance during the final hours. The fact that they were unaware of this deteriorated performance and were even in fact, congratulating themselves on their superior performance may have ominous implications for space flights involving acute fatigue stress.*[8]

★

While human experiment volunteers like Simons and Stapp gained both public attention and professional status because of their daring achievements, space pioneers in the Soviet Union worked in relative anonymity. Fame was not an option even for the Chief Designer, who was most responsible for his country's space success. Although his name had previously appeared in a few public documents and scientific papers, after *Sputnik* it was literally erased from these sources. The secrecy was later explained as a matter of security. It protected both the Soviet space program and the man himself. But in refusing to acknowledge the leader at Baykonur, the Soviet bureaucracy was also making a political point. All achievements belonged ultimately to the state, and service to the state was its own reward.

Nothing in the historical record suggests that the Chief Designer was at all concerned about this forced anonymity. Nor was he eager for public acclaim. He was motivated by the challenge of space, and after meeting with Khrushchev he began working furiously to launch the world's most powerful rocket on yet another breakthrough mission. He summoned half a dozen top engineers who were away on vacation — their reward for the success of *Sputnik* — back to work.

In less than a month they would design and build a satellite that would protect a small dog through launch and provide it with food, water, oxygen, and a stable temperature. The mission would provide scientific data on the pup's vital signs via surgically implanted sensors that would send signals to a radio transmitter. But the dog's primary contribution to Soviet society would be political. At home, the Chief Designer's countrymen would find pride and reassurance in a second great success in space. Around the world, and most especially in the United States, people would again be forced to recognize the USSR as a great power.

To be sure they were prepared for the precise message he wanted to send, Khrushchev invited a representative for the most important newspaper in the West to interview him. James "Scotty" Reston, Ford Fairlane driver and Washington bureau chief for *The New York Times*, was well aware of Khrushchev's dark side, but also seemed to favor listening to the Soviets over Cold War isolationism. Having reported on World War II, he had been impressed by the military will shown by the USSR. He had also been chastened by the talk of preemptive nuclear war, which he had heard at the highest levels in the American Department of State.

On October 7, Reston, a youthful-looking forty-eight, dressed in a dark suit, bow tie, and white shirt, arrived at Khrushchev's office at the Kremlin at 3 P.M. At first Khrushchev relied on an interpreter, who spoke in a kind of propagandese and refused to elaborate or clarify Khrushchev's statements. Reston objected and Khrushchev relented, turning the encounter into a three-hours-plus dialogue on war and peace and communism's gifts to humanity.

Khrushchev spoke in a matter-of-fact way about *Sputnik*, telling Reston that on the night of the big event, "I congratulated the entire group of engineers and technicians on this outstanding achievement and calmly went to bed." But he clearly recognized the satellite's propaganda value. Since he couldn't match America's spending on weapons such as bombers and fighter jets, the Soviet leader's military strategy depended on making his enemy worry about a new one that he claimed would make conventional aircraft obsolete.

"We have all the rockets we need, long-range rockets, intermediate-range rockets and intercontinental missiles," he said, as if the R-7 *Sputnik*-launcher was an actual deployed weapon. "The United States does not have an intercontinental ballistic missile, otherwise it would have easily launched a satellite of its own. We can launch satellites because we have a carrier for them . . . such are the actual facts."

The implication was clear. In light of *Sputnik*, war fought by intercontinental rockets carrying atomic warheads was technically possible and only the Soviets were properly armed. What Khrushchev failed to mention was that the USSR had no stockpile of these missiles and was capable only of launching them one by one. And given the difficulties of fueling these machines with liquid propellant, the chore would take at least a day to accomplish each time. Even then, given the state of the technology, many of the rockets would fail.

Some of these facts were known by U.S. officials, who also understood that America's own major missile projects—six in all distributed among the Army, Air Force, and Navy—were likely to produce more efficient, reliable, and accurate bomb-throwers than the R-7. But they couldn't call Khrushchev's bluff without revealing the existence of the U-2 and other secret resources. Their response was also affected by an important subtext in the chairman's remarks. Even as he flexed his missiles, Khrushchev spoke about the need for international safeguards on rockets and satellites. He flattered Reston, noting how "you are held in esteem by your readers," and implored him to tell the American public "that our people, our state, want peace."

Just how peace might be achieved remained unclear as the mercurial Khrushchev lurched from statesmanship to propaganda. He lectured the *Times* reporter and through him the American people on the virtues of Marxism and his complaints about capitalists who "do not want disarmament, fearing to lose their profits." Defense contractors "are squeezing gold out of the sweat and blood of the workers," he said. He then added, conspiratorially, "As to how they do this, you probably know yourself."

Khrushchev answered so many questions and spoke at such great length that the *Times* published a lengthy verbatim transcript of the interview. Not included were Reston's descriptions of the encounter, which he wrote years later. The chairman was by turns impatient— drumming a plastic letter opener on the table—and impertinent as he challenge Reston's knowledge of world affairs. He also seemed to care about the impression he was making on his guest, telling Reston that he hoped he had not insulted him. The reporter assured the leader of the world's second nuclear power that his feelings had not been hurt.

★

After Khrushchev made his points with a single, carefully selected news reporter, Dwight Eisenhower prepared to face a roomful of them at the White House. For days other politicians and commentators had been talking about *Sputnik* in cataclysmic terms. Now the president would have his first chance to deal with the whole matter of space, national security, and the meaning of the new satellite. Years later it would be obvious that while the president understood all the military issues, and even the national mood, he misread the political symbolism of the moment.

At the press conference, Eisenhower sounded more like an executive reassuring stockholders than the leader of the free world responding to a crisis. He described how America had invested many millions of dollars in missile development and that "there hasn't been any unnecessary delay" in the effort to launch a satellite. He poured on the grandfatherly charm, joking about how many people had suddenly become space experts and insisting there was no need for a frantic effort to match the Soviets. The coolheaded pursuit of plans already set—including a promised Vanguard satellite launch in December—would satisfy America's scientific and military needs.

Eisenhower's calm came close to matching the actual public mood. Although later generations would assume that a panic had immediately swept the country in the days after *Sputnik* went into orbit, several surveys would show that most people were not terribly shaken up by the first Soviet satellite. Just 13 percent felt their country was dangerously behind the Soviet Union in the exploration of space, and more than 60 percent expected America to claim the next big scientific advance. This steadiness reflected the traditional character of small-town America as reported by writer William Furlong just days before Sputnik flew. After visiting Keokuk, Iowa, Furlong concluded that much of America was marked by its tranquillity and calm isolation. Social clubs flourished, civic pride ran deep, and "the world of H-bombs, disarmament and inflation is very remote," he wrote.

However, as "unshakably constant" as places like Keokuk might be, they did not shape the national political agenda. This power was held by party leaders, the press, and others who interpreted events and had access to the national media. By October 7, these Americans were already suggesting that the White House had neglected its duty while allowing the Soviets to capture strategic high ground. Typical was Eric Sevareid of CBS News, who warned that soon the Soviet Union could "stand astride the world, its military master." Like many, Sevareid also criticized the president for being too focused on balancing the budget and keeping taxes down when deficit spending for defense might have gotten America into space first.

With alarm and even anger in their voices, Ike's critics spoke with passion that was missing from the president's reassurances. Records from those days show nearly universal alarm in the press posed against administration responses that seemed so relaxed as to appear out of touch. It didn't help that as Eisenhower tried to put events in perspective, he sometimes sounded like a very old man describing a very newfangled thing he didn't quite understand:

> Now, quite naturally you will say, "Well the Soviets have gained a great psychological advantage throughout the world" and I think in the political sense that is possibly true. But in the scientific sense it is not true, except for the proof of the one thing; that they have the propellants and the projectors that will put these things in the air.

Propellants? Projectors? Things in the air? Eisenhower's language made him sound fusty and unsophisticated. And while his claim that there was no space race underway jibed with publicly stated administration policy, it wasn't true. A day before, the president had quietly told his defense secretary "to have the Army prepare its Redstone at once as a backup for the Navy Vanguard." Meanwhile the Vanguard group was scrambling to make a December launch. This event was initially planned as a test run, but in light of *Sputnik*, the president announced that the rocket would attempt to place a grapefruit-sized satellite into orbit. If this move didn't signal a race, what would?[9]

★

Vanguard was led by a highly regarded, forty-nine-year-old astronomer named John Hagen, who was the temperamental opposite of his intramural competitors, the dynamic von Braun and the swaggering General Medaris. Long-faced, with thick-rimmed glasses, Hagen was a pipe-smoking eccentric. He drove a tiny old English car to work every day and packed his lunch in a black metal lunch pail. He also insisted there was no competition at all between the Soviet and American space programs. (It had been the good-natured Hagen who played the recording of *Sputnik*'s "voice" at the National Academy of Sciences for Anatoli Blagonravov.) His mild manner and never-ending patience were likely appreciated by his former academic colleagues at Wesleyan and Yale but were no help at all when it came to rallying the Vanguard troops against political neglect and bureaucratic inertia. As James Van Allen would eventually recall, Hagen seemed to have "no real leadership style at all."

On the day when Ike met the press, Hagen was summoned to Capitol Hill to report to the Johnson subcommittee's chief investigator, Edwin L. Weisl. Fifty years later, Hagen's notes for the meeting, handwritten on a legal pad, showed the frustration he generally hid from others. He wrote that the Soviets had apparently made *Sputnik* the top priority for the government agencies involved. In contrast, America's civilian satellite effort seemed arranged to produce an "accumulation of circus stunts" with a minimum investment of time and management attention.

The trouble had been plain to see at the launch site at Cape Canaveral, Florida, which was run by the Air Force and its contractors. While the military services, including the Army group from Redstone Arsenal, had full support for their ongoing missile projects, the civilian-oriented Vanguard group would have to beg for facilities and support. Even when it did get proper attention, the program was sometimes afflicted by the low expectations of Air Force officers in charge of things at the Cape. For example, at the first Vanguard test flight on a rainy night in December 1956, an Air Force official tried to intervene during the countdown, shouting "It's gonna blow up!" and "Cancel! It'll never fly!"

Fortunately for Vanguard, the civilians at the controls remained calm and the countdown continued. At zero the rocket, which was a single-stage test vehicle, took off into the night sky over the Atlantic and flew to an altitude of 126.5 miles. The missile performed well, even releasing a small payload carrying a radio transmitter. The tracking stations set up for Vanguard were able to pick up its signals as it glided down into the sea. For a moment, the Navy-civilian program was a complete success.

The excitement over the first test launch wore off gradually, as the contractors building the rocket and its components began to have trouble meeting deadlines. A second launch was delayed for months, and when this rocket eventually flew, it included a second stage that fired as planned. This test also proved that the Vanguard nose cone and broadcasting antennae were designed properly to withstand the heat generated by friction during the missile's acceleration through the atmosphere.

Greater delays plagued the third test rocket, as the prime contractor, the Glenn L. Martin Company, and engine-maker General Electric both delivered bad equipment. The rocket was dented in several places. Wires and fuel lines were improperly installed. Fuel tanks that were supposed to be spotless contained metal chips. One of the Vanguard group's main leaders, field manager Daniel Mazur, became so frustrated that he told a staff gathering "that lump of garbage will have to return to Martin for rebuilding." The body of the rocket stayed, but a GE engine, possibly contaminated by debris from the tanks, was sent back to the manufacturer because it was too dirty to be cleaned properly at the Cape.

A dominant presence in the Vanguard control room, Mazur was a big man with wavy dark hair who wore glasses with heavy black rims.

During this summer of discontent he tried to resign, only to be talked out of it by Hagen. Mazur then made his wife furious by leaving her alone with children sick with the mumps for days at a stretch. At one point Mazur turned to humor to vent his frustration, sending off a two-line poem to a Martin company manager at its Baltimore area plant via Teletype:

> *Rockets are large, rockets are small,*
> *If U get a good one, give us a call.*

Mazur's ditty reflected long-standing concerns about Martin's regard for Vanguard. Upper managers for the government worried that their prime contractors, who did regular work for the military, seemed blasé about this civilian project. One discouraging sign of Martin's indifference was the fact that the company had assigned its Vanguard designers to an old factory where wild birds made frequent deposits on their paperwork.

The troubles with Martin, and other concerns, were part of the history that Hagen recounted in his meetings on Capitol Hill. Hagen understood that impatient politicians might want to engineer a dramatic response to *Sputnik* by shifting the satellite mission to a ready Army missile. But he worried about all that had been gained by the Vanguard program "going down the drain" if the established satellite plan were abandoned.

If he had any doubts about the options available to Congress and the president, Hagen needed only to glance at the other man summoned to join him at the subcommittee office. Brigadier General Austin W. Betts represented the Army missile program and its main leaders, von Braun and Medaris. The Army had never accepted its loss in the competition to lead the American satellite effort and saw now a new opportunity to compete for space and the money to get there.

★

Ultimately the flow of federal cash would become the focus of much of the American response to *Sputnik*. Until this moment, the president, who had made careful spending the core of his political identity, had been deeply concerned about getting the best results for the fewest dollars. Because of this frugality, Vanguard had never actually received a formal budget authorization. Instead money had been grudgingly paid from an emergency fund. Requests for more money were met by delays, which sometimes lasted months.

But with the post-*Sputnik* uproar in Congress and the press, every thoughtful person with any connection to America's space programs understood that the Cold War had reached a new level and the fight would require enormous expenditures. The big corporations that were already involved—Martin, GE, Chrysler, Convair, Rocketdyne, and even one named Grand Central Rocket—could anticipate more work and much more money. The same was true for ambitious scientists and researchers around the country, who would find government officials suddenly receptive to their proposals. At last there would be backing for both their ideas and their careers.

Men and women who had been part of big research and engineering endeavors in the past could also hope that *Sputnik* would motivate the country for a return to the one-for-all, all-for-one spirit they felt during World War II. Whether it was the Manhattan Project or the crash program to develop the proximity fuse, programs that focused large numbers of people on a single worthy goal could give individuals, institutions, and entire nations an exhilarating sense of purpose. When they ended, many of those involved experienced an intense emotional letdown. Some continued to look for ways to re-create the camaraderie and excitement.

James Van Allen, for example, counted the proximity fuse project and the adventure that came with testing it at sea among the most exhilarating experiences of his life. He and many others who had similar experiences during the war would leap at the chance to feel the excitement again. The difference, of course, would be that this time the conflict would be expressed in competing technologies rather than actual war. Super-weapons would be developed in order to dissuade the enemy from attack. And impressive nonmilitary projects, including space exploration, might persuade the world that one political system was better than the other.

Optimistic as Van Allen was, as he continued his work in the Pacific that autumn he worried that he was missing out. Security rules prohibited him from talking about such things via the ship's radio. Messages he received from colleagues were so vague he couldn't grasp what was happening to his satellite dreams. Weeks would pass before he could get to a port and get more information. And in that time he would spend ten days laid up with a lymphatic infection, caused by a small leg injury, which only gave him more time to worry about being out of touch.

He shouldn't have worried. While Van Allen recovered in the *Glacier's* sick bay, his University of Iowa team was already in touch with the officials at the Jet Propulsion Laboratory in California, who had worked closely with the Army on its rockets. JPL had created the additional stages that gave the Redstone the power to launch a satellite. But its director, William Pickering, was eager to establish the lab as a center for the development of space exploration vehicles, rather than rockets. These machines would have real scientific value, while rockets were merely transportation.

Van Allen had devices for measuring cosmic rays that would be compatible with both the Navy and Army satellites. During the summer one of his students, George Ludwig, had even visited Huntsville, where no one was supposed to be working on a satellite launch, to discuss what might happen in case of a launch. When JPL called in the days after *Sputnik*, Ludwig became part of the rocket underground, quietly readying for an Army launch. Soon he would drop out of classes and move to the Los Angeles area with his wife (pregnant and entering her third trimester) and two daughters to help get Van Allan's cosmic ray counter and its radio transmitter ready to be sent into orbit by Medaris and von Braun. In the event that Vanguard failed, this equipment would ride atop an Army rocket.[10]

Needing a name for the package of instruments that Ludwig was making, the group at JPL, where card playing was a regular pastime, thought about how the Soviets had managed to win the first hand in the long game that would be the space race. Engineer John Small, a satellite development leader, noted that, "The winners laugh and joke while the losers yell, 'Deal!'" By unanimous acclaim, the group decided that for the time being the satellite would be called *Deal*. When the work expanded to include follow-on packages of instruments they were to be called *Deal II*, *Deal III*, and so on.

★

While Van Allen's men worked to make his big idea part of America's response to *Sputnik*, other scientists and military schemers developed proposals for all sorts of experiments. One Air Force group proposed launching a swarm of killer satellites that could detect and destroy enemy missiles. Another suggested a nuclear-powered airplane. Neither of these proposals would get very far, but an equally jarring notion was immediately put on a fast track.

Soon after *Sputnik*, a former elevator engineer and self-trained physicist named Nicholas Christofilos began to design a large experiment to see if a Soviet attack on America from the upper atmosphere could be thwarted with radiation. Although military officials dreamed of more conventional antimissile systems, and had spent millions on devices to shoot down incoming warheads, experts doubted this kind of defense would be practical, or even workable. The latest estimate held that twenty small missiles would have to be fired to stop one incoming warhead. For this reason, the idea of a more esoteric technological fix quickly gained support, especially in the community of physicists that knew Christofilos and respected his energy and imagination.

Christofilos had spent most of his early life in Greece, where he pioneered important theories and practical machinery for studying ions and electrons. With no doctorate in physics and an unusual writing style, his papers on early atom smashers were, at first, rejected by leaders in the field. Scientists at the University of California at Berkeley, for example, considered his writings a mess and dismissed him as a crank. He submitted revisions, which were initially ignored, but when he appeared in person at a federal lab on Long Island to explain his designs, he got a warmer reception. He was paid $10,000 for the work he had done and quickly found a job at a University of California research center where he had previously been dismissed as a crackpot.

By the late 1950s, Christofilos was part of the physics establishment in the United States, working on a variety of academic, government, and industrial projects from his base at the University of California at Berkeley. One day *Life* magazine would dub him "The Crazy Greek," but for the moment he was, in the eyes of his colleagues, one of those rare few who were so brilliant they could work their way into high-level science without a terminal degree.

He got by, in part, on the strength of his personality and political affiliations. A muscular and intellectually intense man with brooding eyes and black, slicked-back hair, Christofilos belonged to the small community of fervently anti-communist hawks associated with Edward Teller, who had led research on the American hydrogen bomb. In 1954, Teller had been alone among atomic experts in testifying against Robert Oppenheimer, the father of the A-bomb, before federal authorities stripped Oppenheimer of his security clearances. (More representative of the physics community was Nobel laureate Isidor Rabi, who said, at the same hearings, "He gave you the atomic bomb. What do you want, mermaids?") Teller's politics separated him from his peers in physics, and to a lesser degree Christofilos was also isolated by his deeply suspicious attitude. Physicist Herbert York, who worked closely with both men for many years, would eventually conclude that they were paranoid. But in the moment after *Sputnik*, Teller and Christofilos found favor with policymakers who were wary of softhearted scientists and eager for a strong response.

Christofilos was highly imaginative and his antisatellite experiment would reflect this quality. He theorized that atomic explosions set off at the edge of space might create a lingering shell of subatomic particles that could burn up warheads bearing down on America. Of course, the only way to test the theory would be to launch some missiles to deliver atomic bombs that would explode a few hundred miles up, and then see what occurred.

At almost any other moment in history Christofilos's idea for making a radioactive missile-stopping shield wouldn't have had a chance. But in the days after *Sputnik*, government officials encouraged him to draft a plan. The time for bold experiments in space had arrived.

Many of the grandest ideas of this time would be pursued in secret. Christofilos, for example, would see his theory tested in the South Seas during a classified exercise code-named Argus. Another secret project, begun by the CIA in *Sputnik*'s wake, would eventually put the first photo surveillance satellite in orbit over the Soviet Union. Conducted entirely off the books so that its budget wouldn't attract attention, Corona wouldn't be made public for decades. Although these programs were conducted in secrecy for obvious strategic reasons, obscurity also protected them from negative publicity over technical failures, policy mistakes, and even the escalating costs charged to the taxpayers, which ran into the billions of dollars.

Other projects, including Vanguard and the Army's rocket program, were pursued in public view in communities that would be transformed by the space race. The Army's fast-growing Redstone Arsenal in Huntsville, Alabama, where the rumble and roar of rocket engine tests shook the Tennessee River Valley, was one of these places. But the real drama and excitement in the American rocket campaign occurred about 700 miles to the southeast, on the Atlantic coast of Florida. There the night skies often blazed with the light from streaking missiles and the horizon in daylight was sometimes filled with the black smoke from failed experiments. Day by day, the very landscape changed as once sleepy towns were filled with newcomers, people with new ideas and new visions of the future.

In the late 1950s no place in America was undergoing faster, more radical change than the area around the 15,000-acre missile test range at Cape Canaveral. Until the 1950s, this region had been one of the more isolated and undeveloped places on the East Coast. Except for the families of a few fishermen and citrus farmers, the Cape itself was a vast plain of scrub trees and palmettos, inhabited mainly by mosquitoes, shore birds, rattlesnakes, and the occasional Florida panther. Even tourism had been slow to develop in this part of Florida, although a few mobsters from the North appreciated the quiet so much that they bought properties in nearby Titusville.

The town nearest the launch facilities, Cocoa Beach, had been home to only a few hundred year-round residents in 1940. It grew a bit, and its quiet mood—ruled by the rhythms of the sea and the sun— began to change during World War II when an air base was built south of town. More new people came as the postwar rocket programs began to share the Cape's facilities, firing their experiments out over the ocean. By the time *Sputnik* went into orbit, locals had begun to understand that the place they knew was being transformed. But they had no idea how radical the change would be. In a very short time, the Cape and Cocoa Beach would become the focus of the world's attention. Technology was about to wash over this barrier island like a hurricane-driven tide.[11]

THREE

The Cape

In a few years, nothing that happened around Cape Canaveral would shock Roger Dobson. It was a wide-open American boom town, where newcomers arrived every day, eager to make big money and to be around something exciting. Soon enough the astronaut celebrities would arrive, bringing with them groupies and glamour. Presidents and movie stars would visit. But in October 1957, the region's transformation from rural isolation to the center of America's space industry was just beginning, and a good ol' boy could still be surprised.

Dobson, tan, square-shouldered, and just twenty-four years old, worked at the just opened Celestial Trailer Park (every new thing was getting a space name), which his stepfather, a commercial fisherman, had built on seven acres of family land. Between 1950 and 1960 the region's population would triple as launching pads and other facilities were built and then went into operation. The demand for housing was so strong that trailers lined both sides of the road during the week before the park opened. The men and women who immediately filled every vacancy at the Celestial weren't like any people Roger had ever met before.

Consider a warm summer day when his main duty was collecting the rent. Dobson worked his way up and down the rows of trailers until an especially friendly young woman invited him inside. A chat. A cold drink. A certain look. In minutes he was having sex with a woman he barely knew.

It happened more than once, with different women, and years later Dobson would guess that they had been bored or angry with husbands who never seemed to come home from their jobs. Dobson found these encounters exciting and dangerous. He would never forget hiding in a closet when someone knocked on a trailer door. To his relief it was not a husband suddenly arriving home, and he was able to escape undiscovered.

"This was before the 1960s, but a lot of these people had that 1960s attitude," recalled Dobson many years later. They also brought a huge amount of cash to the local economy. Missile activities generated a $6 million monthly payroll at a time when, for comparison, commercial fishing was worth less than half that much *per year*. Much of this money occupied the pockets of single young men who eagerly sought ways to spend it. As a result, "things were wide open here," recalled Dobson, "especially at night."

Bars, restaurants, and nightclubs sprang up in Cape Canaveral and in Cocoa Beach to serve workers who needed to blow off steam. Almost any kind of trouble could be had, from high-stakes gambling to drugs. Perhaps the only vice that didn't take root was prostitution. As Dobson and others would recall, it wasn't necessary. At the Cape people lived by the code of special circumstances. Difficult and dangerous work performed under extraordinary pressure in a foreign environment loosened the rules. Anyone who developed Sunday morning remorse could attend a Roman Catholic Mass at a local bowling alley, where Communion was offered beneath a sign that read "Bowl Away Your Cares and Troubles." The bowling alley was owned by an ambitious local businessman named Jake Brodsky, and some affectionately called the place St. Jake's-By-the-Sea. Brodsky also owned a popular motel with a bar, which meant he was involved in both ends of the sin-and-atonement cycle.

Wild nightlife came naturally to outsiders who considered the palm trees, balmy weather, and warm ocean waters and felt like they were in a vacation paradise. It also helped to reduce the stress that came with living in a place that could feel achingly lonesome. Tropical flowers and jumping dolphins could not replace the longtime friends, extended family, and community connections the space migrants had left behind. They did nothing to ease the challenge of daily life in a strange new place that was not ready to accept so many immigrants.

In reality, very little, except perhaps sex and alcohol, came easy to the space itinerants. Newcomers arrived in such numbers that the region needed one new schoolroom per week to accommodate their kids. Housing was so tight that some laborers slept in the concrete drainage pipes that were waiting to be installed at construction sites. (On nights when the mosquitoes and watchmen weren't too thick, these pipes were as cozy as cocoons.) Others tripled up in motels—the new ones were named the Sea Missile, the Starlite, and the Vanguard—where three-quarters of the rooms were booked on a long-term basis by missile contractors, including Convair, North American Aviation, Bell Telephone, and AC Spark Plug. Among the available apartments, many came without potable water because the municipal system wasn't fully built and many wells in the area produced only cloudy, foul-smelling stuff. With land prices rising fivefold in seven years few newcomers could afford to build new homes.[1]

★

Along with the young adventurers who sought as much fun as fortune in the space boom, the Cape area attracted hundreds and then thousands of career military officers and seasoned civilian engineers. These upper-class migrants, who came with their families and furniture, swelled enrollment in public schools, taxed public services, and put even more pressure on the local real estate market.

The lucky few were able to buy or rent existing homes close to the missile test range. Others put down payments on houses being constructed in new subdivisions, some of which were built on land created where the marshy riverside had been filled in with rock and earth. While they waited, these families crowded into rented rooms where they prayed that the contractor would complete the job on time. In some cases a handshake deal was fulfilled on schedule as bargained. In others, delays and cost overruns led to huge motel bills and lawsuits.

Families who weren't willing to wait for a new house and could find nothing closer to the test range settled in more distant towns, such as Rockledge to the west and Titusville to the north. In these places they encountered a mixed welcome from locals who appreciated the money that the space program poured into the local economy, but were wary of the influx of so many strangers, especially Yankees.

Space immigrant families also had to cope with the unattached young workers who didn't have Roger Dobson's trailer park luck. Some of these men took to attending local high school football games and hung out at roller rinks, bowling alleys, and drive-ins looking for dates. More than a few found what they were seeking and weddings involving eighteen-year-old brides and twenty-something Cape laborers became common.

Determined to protect their beautiful seventeen-year-old daughter, JoAnn, Donald and LaVerne Hardin declared space workers off-limits soon after they arrived in the Cape area community of Mims, just north of Titusville, in December 1956. The Hardins must have felt a bit like pioneers in the Old West when they brought three children, a dog, and some cats to Mims from the genteel town of Jacksonville, Alabama, home of a state teachers college. In Jacksonville they had lived in a big, graceful old home tended by a housekeeper. The children had attended a special university-run Laboratory School and their afternoons were devoted to tennis and music lessons. JoAnn, in particular, had thrived in Jacksonville, developing particular passions for math and science. Marie Curie became her heroine.

In Florida, JoAnn and her siblings attended schools that were obviously inferior to the one they had left behind. The nearest music teacher was forty-five minutes away and their neighbors, some of whom proudly called themselves "crackers," were far more likely to spend the afternoon in the swamp shooting game for the dinner table than hitting fuzzy white tennis balls.

For more sophisticated newcomers like the Hardins, relief from the culture shock of rural Florida came when they spent time with friends and colleagues who worked for the Army Redstone project. Like the Hardins, many of the higher-ranked Cape settlers had been affiliated with the missile facility in Huntsville and had accepted transfers to the

Cape area. In their new homes they tried to create some version of middle-class normality between the honky-tonk and the swamp. They cherished the dinner parties and other events when they could come together and relax.

Weeks after she settled in her new home, LaVerne Hardin eagerly dressed her children for one of these encounters, a Christmas party for the Army Ballistic Missile Agency. Several of the project's German scientists, the most cultured and educated people in the community, would attend, and, the children hoped, they might be joined by St. Nick. The banquet room at the Cocoa Beach hotel was crowded when the Hardins arrived and Christmas spirits were flowing.

When the big moment came and the crowd focused its attention on a special arrival, LaVerne suddenly turned to her daughters and blurted, "Cover your eyes girls!" In boom town fashion, someone had ordered a stripper for the party instead of a Santa.[2]

★

The Christmas stripper would enter Hardin family lore along with tales of LaVerne's struggle for normalcy in a place without proper supermarkets or a department store. Hardin, who had worked in Alabama, would soon realize she wanted more than the life of a homemaker and take a job at the Cape. One of few women to get a job on the test range, she was a clerk at Hangar D, where crews put together the latest generation of Army rockets based on the Redstone design. Like so many in the Army team, LaVerne was intensely loyal to Wernher von Braun and General Medaris. She believed that they, not the Air Force, Navy, or civilians, were America's best bet to reach space.

When he was alone with his group, the self-confident Medaris didn't hide his disdain for the others who sought to take America into space. He hated the Air Force, insisting that as a new military service it suffered from an "inferiority complex" and should have never been separated from the Army. Medaris was also certain that the Navy and Vanguard had never deserved the satellite job. Eisenhower was ultimately responsible for this decision, and Medaris wasn't favorably disposed toward him, either. He opposed the whole idea of putting "a soldier in the White House" because Ike was too impressed with his own judgment when it came to defense and not open to the counsel of "today's military professionals." (Not surprisingly, Medaris was not the president's favorite, either. Ike considered him to be an overly pushy, supersalesman type.)

The general believed, in contrast with the president, that space combat was inevitable because "wherever man has gone . . . sooner or later he's managed to get into a fight." He favored von Braun's ideas for militarizing space with spy satellites and an orbiting space station that would allow, as he put it, "domination of the whole planet."

Medaris's loyal workers shared many of his attitudes and felt both frustrated over the limits placed on their efforts and competitive with the other missile programs. When the others scheduled test flights, the crew in Hangar D, including LaVerne Hardin, would take to the roof of their office building to cheer against them. For example, as the Air Force prepared to launch its Navaho missile, they chanted, "Navaho will never go! Navaho will never go!" It rarely did go, failing to fly in ten of its first eleven attempts.

Captivated by the sense of mission and the spirit of her co-workers, LaVerne loved working in a place where the very air vibrated with the roar of rocket engines. She was on the front lines of the Cold War, and the stakes in play included the nation's security and honor. The mission wasn't so different from, say, the storied Manhattan Project and the mood at the test site was marked by the same kind of urgency, camaraderie, and sacrifice. The scale of activities at the test range was enormous. Construction workers were constantly clearing land, building roads, and pouring concrete for launch pads along Missile Row, the line of launch facilities that flanked the ocean. Trucks hauled in rockets and other equipment, and when a launch was scheduled the tension could be felt by everyone.

But as much as she relished the work, LaVerne was concerned about her children, who had to accept that their parents worked long hours in a potentially dangerous setting. The technology they commanded was still new, and the pace of their work was governed by urgent geopolitical factors. If the Soviets had a rocket that could deliver an atomic bomb across the seas, for example, America must have one too, and as soon as possible. For these reasons, space workers and their families had to accept certain risks and hardships.[3]

★

The daily difficulties that Cape workers faced began with the challenge of simply getting to their jobs on time. As the test range grew to include more than a dozen launch pads, assembly hangars for rockets, and an industrial area for contractors, commuting became a daily contest. None of the local highways was big enough to handle the traffic. For example, the best road in Cocoa Beach, A1A, was just two lanes and it was crowned so steeply that any lapse in concentration could send a driver drifting to the right and into deep sand that could be impossible to escape. One spinout or breakdown and all traffic stopped while a tow truck was summoned. Sometimes the line of cars would back up almost to Patrick Air Force Base, which was nearly twenty miles south of the test range.[4]

Once a commuter got to work, he or she encountered more uncertainty. Overtime was the norm, but on some occasions a worker might be forced to stay on site for days on end. These marathon shifts occurred when a missile launch was set but the countdown was suspended because of weather or faulty equipment. Often the count would be resumed only to stop again when another problem arose.

Those waiting ate military rations and, if they weren't assigned to some task, fought against boredom any way they could. Daredevils raced their cars on empty Cape roadways. Fishermen improvised tackle out of twine and shiny bits of metal to catch bluefish schooling close to shore. (Freezers intended for battery storage preserved the catch.) Barred from bringing firearms to work, hunters considered the game that roamed the Cape and grew as frustrated as house cats watching birds through a window. They eventually made a monster-sized, compressed-gas-powered shotgun out of a three-inch metal tube, which they placed on a flatbed truck. When a flock of ducks settled onto one of the many ponds on the range, and bored missile men were free to respond, they climbed into the truck and headed for their prey like soldiers with a jeep-mounted machine gun. A single shot from the compressed air cannon—roughly ten pounds of number six nuts—would kill a dozen or more birds at a time. The men would wade out to collect the ducks, and then distribute them to the gunnery crew and the security guards who were supposed to police this kind of play.

When countdowns were halted at night, passing the time was more difficult, especially for the unlucky workers who had to sleep in tents. They might play cards or talk by camp light, but a lantern's glow tended to attract mosquitoes. The Cape was an ideal breeding ground—plenty of still water in shallow ponds and swamps—and the bloodsuckers were so thick that the only effective way to suppress them was the regular application of DDT. Trucks spraying this pesticide even drove into hangars to keep the insects under control. So many mosquitoes were killed in hangars that push brooms would be needed to sweep them away. Red streaks on the floor betrayed the last meals of the condemned. Through all the hardship and the waiting, workers tried to keep alert because eventually most rockets were fired and at the moment of ignition you might have to run for your life.[5]

In the seven years since the first launch on July 24, 1950—technicians destroyed that particular rocket, which was called Bumper 8, when it veered off-course—well over 200 missiles had been ignited on the concrete pads at the Cape, with varying degrees of success. These rockets, including Larks, Snarks, and something called a Bull Goose, were all designated "test vehicles," which meant they were custom-made and intended to advance basic technologies. When they worked, local civilians might notice a rumbling roar followed by a white plume streaking into the blue sky or a flame rising in the middle of the dark night sky. Of course, many missiles quietly refused to cooperate and just sat dumbly on the pad, failing to obey commands. Others malfunctioned once their engines were fired and either crashed, exploded, or were intentionally destroyed in midair.

Activities at the test range, especially the many failed experiments, were generally kept secret. But it wasn't hard to figure out what was going on. At some launch centers, crews turned on red, pre-firing warning lights attached to poles near the pads. Local commercial fishermen—shrimp was their main catch—knew a launch was set, and that they should steer clear of certain offshore areas, whenever a red ball was hoisted to the top of a signal pole at Port Canaveral. Less savvy sport fishermen might be buzzed by small aircraft sent to drive them away from certain areas.

In nearby communities, waitresses and hotel clerks understood that when certain out-of-town engineers or military officers appeared, a rocket was about to go up. Kids who paid attention to their parents' work schedules developed a sense of when a launch was imminent. Others listened for clues in adult conversations and a few seemed to get direct briefings from their dads. At Titusville High School, a redheaded boy named Dalton Cairnes, whose father worked for rocket-maker Chrysler, always seemed to know when to organize a rocket watching beach party at Jaycee Beach, where civilians could actually see some of the missiles waiting to blast off. Army launches were frequently made on Friday nights, which meant that Dalton and his friends could stay out late in hopes of seeing something exciting.

The scene at a place like Jaycee Beach, perhaps the best spot in the world for casual rocket-watching, offered a strange blend of futuristic technology and timeless wilderness. Flanked by the Cape's empty acres of sand, palmettos, and sea oats, the beach had its share of wildlife, including a large, resident indigo snake (the largest species in North America, they reach up to nine feet) that enjoyed wrapping its blue-black body around the cool stones of a well. Soon enough the sons and daughters of missile pioneers would consider it almost routine to be greeted by this snake and other animals as they laid out blankets and waited to see either a successful flight or an earth-trembling accident.[6]

It was impossible for workers at the Cape to keep secret the big blowups that occurred with regularity. The fuels used for rockets were so volatile that explosions sent balls of flame a hundred feet or more into the air and created clouds of smoke that would be visible for twenty miles on a clear day.

Sometimes errant rockets came crashing down outside the 15,000 empty acres of the test center's reserve. One early rocket veered sharply south after takeoff, passed the Cape Canaveral lighthouse (a 160-foot-high tower painted in broad stripes of black and white), and quickly escaped the range of the radars that were set up to control it. It crashed fifty miles away beside a popular restaurant in Vero Beach called the Ocean Grill. No one was hurt but a few windows at the grill were blown out.[7]

The Vero Beach crash was just one example of the risks involved in every launch. It also illustrated the government's limited ability to protect the public from these strays. Not that the feds didn't try. All the sites considered in the late 1940s, when the space center was being established, were adjacent to open water where missiles might fall safely to rest. Officials had decided that no place was better suited to rocketry than Cape Canaveral, and not just because of the Atlantic Ocean. The British were willing to let the Americans build a downrange tracking station in the Bahamas. And because rockets could be fired toward the east, attempts to orbit satellites from the Cape would get a little boost from the rotation of the Earth.[8]

Of course, all the site planning did nothing to protect the crews at the launch pads, who faced the greatest danger of all. For example, in 1953 a Bomarc missile seemed to take off normally, but then went out of control, flying in circles over the Cape before returning to Earth a few hundred yards from where it had been launched. The event was documented on a reel of film that showed the rocket's looping flight and then ended with the cameraman running for his life when he realized the rocket was headed straight for him. Emergency officers who rushed to put out the fire started by the Bomarc were shocked to discover that a wooden shed that had been ablaze was filled with smaller rockets and their highly explosive fuel. In a similar incident, a missile designed to home in on the iron and steel hulls of warships actually reversed course and came back to the Cape to destroy an old U.S. Army tank that sat next to a launch pad and had served as a makeshift blockhouse.

Fortunately, few rockets turned back, cartoon style, and sent men scampering before they hit the launching area. And in only one case did an errant rocket fired from the Cape make it all the way to another country. In 1956 a winged missile called a Snark was tracked beyond the southern horizon as it refused to obey commands to turn or self-destruct. Last seen by American radar as it entered Venezuelan airspace, its remains would eventually be discovered by a Brazilian farmer.[9]

More commonly, failed launches resulted in midair explosions over the ocean or the Cape itself. Quite a few of these blowups occurred before a rocket even left the vicinity of the launch complex. In October 1956 two poorly soldered connections inside an Army Jupiter A (a Redstone rocket named for the king of the Roman gods) came loose seconds after liftoff, sending the missile out of control. As it headed for a nearby launch complex—two pads, a blockhouse, and other facilities—being constructed for the Air Force Thor, the range safety officer hit the self-destruct button and blew the Jupiter A and its nearly full fuel tanks to pieces. For a while people at the Cape joked about the Army attacking the Air Force with an IPM—Inter-Pad Missile.

When rockets exploded over the test range, flaming bits of metal would fall from the sky like fiery hail, sending observers and ground personnel running to hide under cars and trucks or inside concrete buildings. Hot pieces that landed in palmetto thickets or groves of scrub trees could start fires in many different spots. Crews from the test range fire department, who were on alert for every launch, searched for the center of a brushfire by watching for frightened animals—rattlesnakes, birds, rabbits, raccoons, wild boar—and headed for the spot they seemed to be fleeing.

A big fire had burned on a launch pad at the Cape the day before the Soviets launched *Sputnik* when a rocket called Thor 107 committed the equivalent of missile suicide. Sixty-five feet tall and capable of producing 100,000 pounds of thrust, the Air Force's new missile was comparable to the Army's latest generation Redstone rocket (the Jupiter C) and was designed to carry nuclear warheads up to 2,000 miles. Four of the five previous tests had ended with destroyed rockets. This one turned out worse than the others, with the rocket barely getting off the ground before it fell and collapsed in on itself in a heap of fire and smoke.

Although designers and technicians considered such losses normal, outsiders might consider Thor's record and conclude that the program was a mess. This would have been difficult, however, since the details of the Air Force's effort to create an IRBM (intermediate range ballistic missile) were kept secret and tests didn't receive public scrutiny. On the day Thor 107 died, the Air Force didn't even make an official comment.

A brief announcement—"a missile was fired"—was made the following week when Thor finally performed as planned. The sixty-five-foot rocket took off a little after 11:30 A.M. on Friday, October 11. People on the nearby beaches reported it flew straight "without any wobbling" and was out of sight within a minute. Although the military didn't actually confirm that the rocket was a Thor, the press and stock market knew it almost immediately. A down day at the New York Stock Exchange suddenly turned upbeat after lunch, with Douglas Aircraft, Thor's prime contractor, leading a big rally on the busiest trading day in more than two years.[10]

★

Unlike Thor, the Vanguard program, which carried America's satellite dreams, did not enjoy the cover of secrecy. Reporters examined every stage of the project's development and were actually permitted to watch tests. When two successful flights were made in late 1956 and early 1957, the press was universally positive. But then came months of delays and reporters began to use the phrase "ill-fated" as they described Vanguard, a practice that irritated space workers like John Neilon almost as much as the delays themselves.

Neilon was an early Vanguard team member who had moved to Cocoa Beach in the spring of 1957 to work on the radar that would track the missile. At the start of the assignment he sensed that the program was a poor relation on the Cape. He was right. When requests were first made to base Vanguard at the test range, Air Force officials, who managed the launch site, didn't like the idea of adding a purely scientific program to a site devoted to urgent, secret military projects.

Even after they were admitted, the civilians had to fight for support from the groups that managed hangars and launching pads. This second-class feeling extended to a very personal level in April 1957 when the federal payment sent to compensate civilian workers for the high cost of living at the Cape was cut from $12 a day to $8. Months later, even as they were struggling to meet federal deadlines, Vanguard managers learned that they could no longer pay overtime wages.[11]

The competition over equipment and bragging rights consumed valuable time and attention. Conflicts could arise over the most trivial things. Air Force managers complained, for example, when construction of the Vanguard blockhouse, which looked like a miniature Mayan temple, went forward a little faster than the one being built for the Thor.

They were especially irked when a contractor gave Vanguard the first blast-proof window shipped to the construction area.

Little victories like the window were rare for the underfunded Vanguard group, which was subject to competitive sniping from the people working on military missiles, which always had priority. Critics called their program "Project Rearguard," and fed rumors about the Army's rocket experts taking over the satellite mission.

"Initially we had been led to believe that we were the only ones in the race," recalled Neilon. "But von Braun was always playing politics, working behind the scenes, and he never let people forget that he had a rocket that could do the job."

While von Braun and Medaris kept themselves out of public view in the days after *Sputnik*, their higher-ups in the Army couldn't resist complaining. Two generals attending an International Geophysical Year conference in Barcelona repeated the charge that the von Braun team could have orbited a rocket two years ago, and then presented a forecast for America's future in space, including manned missions, direct from von Braun's playbook. The generals left no doubt that the Army was ready to match the Soviets if the Navy and its civilian partners could not.

The Vanguard group came under more pressure when the president promised that a scheduled December launch, originally intended to be another test, would instead be a serious attempt to place a satellite in orbit. The president's commitment made that ambition official, and public.

"But the truth was, we hadn't even fired a single real [multistage] Vanguard," recalled Neilon. "The first test had really been a [single-stage] Viking with a nose cone and the second only had two stages." (Viking was an early model Navy rocket.) A third successful test involved only a first-stage motor. No one in the program knew how an actual Vanguard, with all three stages, would fly. "We always seemed to be doing something for the first time," explained Neilon.

Nevertheless, managers dove into the task of preparing for a satellite launch in December. An aluminum ball the size of a grapefruit was manufactured at the Naval Research Laboratory in Washington, D.C. (The special ultra-clean lab where it was made was nicknamed "The

Ballroom.") It was equipped with batteries, solar cells, and two radio transmitters to signal the Earth from above. At various radio field stations like the one operated by Chester Cunningham in Peru, work on Minitrack, the tracking network, was accelerated so that Vanguard's orbit could be documented.

When the missile that would carry the grapefruit eventually arrived at the Cape in mid-October, a crack was discovered in one of its engines. A contractor quickly sent a replacement and workers at the Cape's Hangar S, the one assigned to Vanguard, installed it. By November, Test Vehicle-3, as it was known, would be ready for the four-and-a-half-mile trip to Launch Complex 18-A on Missile Row, where it would be propped by a steel girder work tower that would support installation of its payload.

But even as they hurried to meet the president's deadline and voiced public optimism, Vanguard's leaders couldn't be sure whether they would succeed. Indeed, given the odds assumed by virtually everyone who worked in rocket development since 1926, when Robert Goddard launched the world's first modern rocket from a Massachusetts cabbage patch, they had no better than a 50-50.[12]

★

In the Soviet Union the same odds confronted the Chief Designer as he considered Chairman Khrushchev's request for a second satellite to impress Mao and irritate the Americans. Rewarded with five days' rest at Prime Minister Nikolai Bulganin's dacha in the subtropical Black Sea resort Sochi, the closest thing to Cocoa Beach in the USSR, the Chief Designer was back at work. He would one day recall this period as the most exciting month of his career, as he led his engineers, scientists, mechanics, and laborers on a race to meet the chairman's November 7 deadline.

Unlike his counterparts at Cape Canaveral, the USSR's lead rocket man did not have to worry about public opinion or rival missile programs. No Medaris or von Braun prowled the windswept Baykonur space center, ready to supplant him. The Chief Designer was also free from public scrutiny. Outsiders couldn't get even a peek at his launch pads, and except for a handful of subsistence farmers, no one lived close enough to see or hear either successful launches or explosive failures. In this isolation, the Chief Designer could take risks that would be unacceptable for the Americans.

Freed from restraints and empowered by Khrushchev, the Chief Designer's first steps in making his second race for space included suspending normal reviews of engineering plans. Instead, drawings would go from drafting tables straight to the workers who would make the components of the rocket and its satellite. Designers would be ruled not by their supervisors, but by their own standards.

Work on the doghouse that would ride atop the R-7 began on October 10, the day *The New York Times* published President Eisenhower's press conference claim that there was no race to space underway. The capsule would be based on a type of vehicle that had already been used to sling dogs on short suborbital missions. But the basic design, for a container that resembled an aluminum beer keg, would have to be upgraded to allow its passenger to survive for days or even weeks.

The add-ons included an automatic feeding system, an oxygen supply and regulators, and a chemical system to remove the water and carbon dioxide generated by the dog's breathing. A cooling fan, which would start to whirl whenever the temperature inside the vehicle exceeded seventy degrees Fahrenheit, was added. Engineers also attached extra scientific instruments to the capsule, including devices to detect cosmic rays and other forms of radiation. All this equipment would push the weight of *Sputnik II* to over 1,000 pounds, six times the weight of the silver ball launched on October 4. But the R-7, which was being quickly assembled at a factory complex on the Baltic, would have more than enough power to throw this weight, and a dog passenger.

The dog was selected with care. Candidates were drawn from a kennel at the space medicine institute run by the Soviet air force. This group was winnowed to three small smart pups that underwent about twenty days of training, which included rides in a centrifuge to simulate the g-force of a rocket launch and confinement in smaller and smaller cages. More doctors than dogs participated in the training. Their goal was to make sure that the first creature in space would not only be cute and furry but also calm and cooperative.[13]

★

Shielded from public view, the teams that rushed to prepare the dog, the satellite, and the rocket for *Sputnik II* surely knew they were preparing a second, not-so-pleasant surprise for an adversary—America—that was not at all fond of being startled by big events. For their part, the Americans continued to monitor the first *Sputnik*, which mocked them with its beep-beep as it passed overhead.

Fortunately for the American ego, *Sputnik*'s battery power was limited. As days went by the signal grew weaker until, after more than 500 orbits, only the most sensitive receivers could pick up the sound. On October 25, the RCA group on Long Island heard nothing when the ball flew by just before 6 P.M. Similar reports came from other outposts. At the super-sensitive listening post outside Lima, Peru, Chester Cunningham's crew detected only the faintest two-tone signal. Soon experts would agree that the little traveler had beeped its last.

Some U.S. experts were sorry that *Sputnik* lost its voice. J. Allen Hynek and Fred Whipple, for example, had figured out how to make detailed studies of *Sputnik*'s path, especially changes in altitude, as it circled the globe from west to east. The orbits showed that the Earth might be more severely bulged at its center than previously thought, and that variations in gravity could draw the satellite closer to the planet's surface at the poles.

Remarkably, most of the American public had quickly adjusted to the idea of a foreign object in the sky. Within days of the launch the baseball World Series—the Milwaukee Braves beat the New York Yankees—replaced the space race as the lead subject on the front pages of many American newspapers. Surveys of public opinion in Europe found *Sputnik* had little or no effect on the status of the Cold War competition. Indeed, people were so calm about the satellite that after a trip through the American Midwest, J. Allen Hynek said he had been "shocked" by the public's complacency. Citizens of the heartland, taking their cue from their president, who played or practiced golf fifteen times in the month of October, were not terribly bothered.

The calm that prevailed in Middle America contrasted starkly with the response among two distinct groups. The first included experts of all types who behaved as if the nation was behind not only in a race for space, but in a contest over the very future of the country. They worried aloud about everything from a rock 'n' roll "craze" that heralded "mental decline," to the erosion of American manhood. Days after the first *Sputnik* flew, several psychologists even warned that thanks to over-bearing mothers and absent fathers "a generation of sissy boys is in the making." The second group of Americans who responded as if the nation were in crisis included more than a thousand amateur astronomers who had signed up for Hynek's IGY sky-watching corps. With *Sputnik*'s sudden appearance many of these observers felt an urgent duty to serve their country and took to their telescopes with patriotic vigor. They were joined by countless Americans who, though less familiar with the stars and planets, began to think and more importantly feel differently about what might be going on in the skies overhead.[14]

One night shortly after *Sputnik* was launched, fourteen-year-old Ralph Stephens was lacing up his roller skates at Scottie's Fun Fair on the causeway to Cocoa Beach when he happened to look up to see a pinkish red object in the northern sky flashing two bright lights. Ralph watched this object for several minutes. It traveled in a straight line, then circles, and then climbed higher. At the highest point of this strange craft's flight, Ralph thought it looked like shades were lowered over its glowing fuselage. Perplexed, he called to some friends and asked them to look up, into the night.

"You're nuts," was their initial response. But then they looked up, caught sight of the lights in the sky, and agreed that they seemed to be moving in a strange way, up and down and then in circles. A state trooper who happened to drive past the roller rink stopped, joined them, and saw that something unusual was going on above. Other witnesses included a police officer in the nearby city of Cocoa and a few citizens, who called the station to report the lights. When the local paper investigated, the public information officer at Patrick Air Force Base said, "Nothing we would be firing would look like that."

When the light seemed to return three days later, the Air Force story changed. According to a printed statement, "equipment checks" conducted by incoming aircraft had been responsible for the two evening displays of lights and action.

In the weeks after *Sputnik* went into orbit, UFO sightings were reported across the country. Some post-*Sputnik* UFO sightings probably involved the satellite itself, and the large rocket booster that shared its orbit. Most UFOs were found to be aircraft, or natural phenomena such as meteors, lightning, or planets reflecting light. A few prompted full investigations. The Air Force paid special attention to reports of a glowing, egg-shaped object that touched the ground first near Lubbock, Texas, and then returned days later to nearby areas of New Mexico.

The Texas encounter was described in the most vivid and consistent detail by more than a dozen unrelated witnesses. Pedro Saucedo, a thirty-year-old farmworker, said the brightly lit object made his truck stall and caused its electric power to fail as it passed overhead. Saucedo said he jumped out and threw himself on the ground. His truck shook as the UFO made a thunderous noise. Saucedo got to his feet once the thing was past, and watched it disappear over the horizon. A nearly identical story was told by a truck driver named Ronald Martin, who said his motor died and lights went off when "a big ball of fire dropped on the highway." Others reported that the object was roughly 200 feet long and glowed with many different-colored lights.

A repeat of the Texas sighting—multiple witnesses, stalled vehicles, bright lights—came a day later near defense and atomic weapons facilities in New Mexico. Two different military patrols at the White Sands missile test range were among many who said they saw the craft. A representative of the Air Defense Missile Command in Colorado flew in to check things out. The sightings were officially dismissed as unfounded and were largely forgotten until they were re-created by Steven Spielberg in his 1977 classic film about UFOs, *Close Encounters of the Third Kind*.

In the post-*Sputnik* era, UFO sightings would continue, with roughly 1,000 reports made to the Air Force in 1958 alone. UFO reports were of interest to psychologist C. G. Jung, who was probably the century's greatest interpreter of how the mythic and the supernatural affected people. Jung had made a thorough and open-minded study of recent books, articles, and first-person accounts related to so-called flying saucers. The outbreak of sightings that followed Sputnik would lead him to conclude that this phenomenon was a product of the human heart and mind rather than extraterrestrial powers.

Because so many who saw UFOs believed they were flown by superior beings who offered the Earth new wisdom, Jung said they represented a new "savior myth" for people whose "minds are cornered by the bad situation in which our world is today." These "perfectly honorable people," added Jung, "are in need of answers to their anxious questions, answers which nobody can give them."

Jung's questions, about the future of humanity in the age of rockets and atomic weapons, would only become more urgent as the high priests of the Cold War—men like the Chief Designer and the leaders of America's space program—accelerated their efforts to create ever more impressive technologies. Eventually the U.S. government would recognize that space exploration in particular might challenge earthlings to adjust to unsettling new realities. To prepare, they commissioned a thorough study of the social implications of discoveries in space, including the possibility that intelligent life might be discovered. The report, aided by experts including the famous anthropologist Margaret Mead, would suggest that mankind was ill prepared for the shock of receiving radio communications from the cosmos or stumbling across artifacts left behind by alien visitors to the Moon. In a worst-case scenario, the experts warned, contact with a race of superior beings from outer space might cause all of human society to break apart.

"Anthropological files contain many examples of societies sure of their place in the universe, which have disintegrated when they had to associate with previously unfamiliar societies espousing different ideas and different life ways," the report explained. "Others that survived such an experience usually did so by paying the price of changes in values and attitudes and behavior."[15]

FOUR

And a Dog Shall Lead Them

Small and a bit shaggy, Kudryavka (Russian for "Little Curly") became a radio star on the night of October 27, 1957, barking into a microphone while a Soviet announcer explained how much she enjoyed training for her upcoming mission. Described as an obedient mutt, she spoke on a national broadcast to listeners in crowded city apartment blocks, small town neighborhoods, and weathered farmhouses. The program also featured one of her keepers, Professor Alexei Pokrovsky, who explained that Kudryavka was going to be the first Earth creature to orbit in space.

"We're working to bring near the time when human travel in space will be a reality," said Pokrovsky, "when people in spaceships will be able to establish contact with other distant, hitherto unknown worlds."

Professor Pokrovsky was interviewed at the laboratory where Kudryavka was being prepared to fulfill her destiny on behalf of her state and her planet. Many a canine comrade had gone before her on suborbital rocket flights. In fact, two weeks earlier a dog that looked a little like a small Siberian husky had ridden a Soviet rocket (one smaller than the R-7) to an altitude of 130 miles and then parachuted safely to Earth, or so the government claimed.

On the broadcast the professor didn't mention whether Kudryavka would make a similar return from her upcoming flight. This was probably wise. There were children listening and as anyone with knowledge of spaceflight knew, Kudryavka's voyage would not include return passage. No one in the USSR, not even the Chief Designer, had figured out how to shield a returning satellite from the heat generated by atmospheric friction. For now, everything that entered space via an R-7 would either continue into the oblivion of the great beyond or return in a ball of fire.[1]

Besides, the mechanics and engineers who were feverishly constructing the satellite had all they could handle with getting the dog *up* safely. The basic equipment for the animal's survival had pushed the weight of the satellite past 1,100 pounds, about six times the heft of the first *Sputnik*. The R-7 had the lifting power to handle this load, but there was no sense in demanding anything more. Still, designers did have some compassion. They added an automatic lethal injection system to the capsule, which could be used to kill the dog before the oxygen ran out.

The euthanasia device and the many other pieces of the vehicle that would be launched from Baykonur—rocket, capsule, instruments—were all shipped from places across the country to arrive by the end of October. Once on site, they were assembled in a hangar near the launch pad. For weeks the Chief Designer managed this work from afar. Then on October 26, the day when Kudryavka barked into a microphone, he boarded an Aeroflot plane in Moscow for a flight to Tashkent, from where he would travel on to the space center.

By the time the Chief Designer reached Baykonur, Kudryavka and her understudy, a dog named Albina, had already undergone preflight surgery to have sensors implanted in their skin to measure their vital signs, including breathing, heart rate, and blood pressure. Other sensors were wired into their foreheads to measure eye movement during both waking and sleeping. These sensors would be connected to a radio transmitter that would send readings back to Soviet scientists. As part of the preparation, Kudryavka, about to become a celebrity, had her name changed to something simpler and more generic. While those who knew her would continue using her nicknames Zhuchka ("Little Bug") and Limonchik ("Little Lemon"), for evermore the world would know her as Laika, Russian for "Barker."

Found on the streets of Moscow, Laika possessed all the qualities that make a mutt a perfect pet. She was intelligent, calm, and easily trained. In shape and color she was a bit like a tiny German shepherd. Her long muzzle was mainly black, as was her face. However, a narrow line of tan fur ran from the tip of her nose, between her eyes, to her forehead. And over each eye, a tan patch the size of a quarter made it look like her brows were always arched in surprise. Gentle but playful, Laika was the kind of dog that children loved. Before her trip to Baykonur, one of her minders had brought her home to see his own children. The time Laika spent playing with them was a reward for the duty to come.

Laika was placed in the padded capsule of *Sputnik II* at midday on October 31. Leather straps kept her from turning around, but she was able to eat some of the special jelled food prepared for her journey. Hours later the satellite, which included a silvery ball similar to the first *Sputnik*, was stacked on top of the R-7. Preparation for the launch went on continuously. The fuel tanks were filled with kerosene and liquid oxygen. The large steel supports that held the R-7 upright were pulled back. Finally at 7:30 A.M. on November 3, the countdown reached zero and the rocket's engines ignited on schedule.

Shaken by ignition and rattled as the rocket lifted off, Laika was subjected to a truly deafening roar as the R-7 climbed into the Kazak sky. She panted furiously and her heartbeat raced to triple its resting rate as the acceleration created pressures several times greater than the force of gravity. The force on Laika's body subsided as the satellite reached weightless orbit, but she had trouble recovering from the stress of the launch. Tests conducted in a centrifuge on Earth had shown that Laika's heartbeat could return to normal soon after excessive g-force was reduced. But in the isolation of the capsule, with no reassuring handlers around, she needed much longer to calm down.

Unlike the first *Sputnik*, which separated from its entire launch vehicle, *Sputnik II* went into orbit with its second stage attached. All this hardware—six tons of stuff worth half a billion rubles—flew along an elliptical path that was about 530 miles high, on average. The rocket-capsule-dog combination circled the Earth once every 104 minutes, passing 160 miles above the orbit of *Sputnik I.*

As far as the experts on the ground could tell, Laika did well on the first two or three orbits. Crude pictures broadcast to Earth by a TV camera in the capsule showed that she was awake and alert. With everything performing as planned, the Chief Designer had given Khrushchev a parade in space to match the one that would soon fill Red Square to mark the fortieth anniversary of the Bolshevik Revolution.

This achievement was an affirmation of power on many levels. First, it showed what the Chief Designer, once a pariah in exile, could accomplish with his government's full support. Second, it demonstrated the capabilities of Soviet industry. Finally, it affirmed Chairman Khrushchev's leadership. Only months earlier, Khrushchev had survived attempts to oust him and finally solidified the power he had begun to acquire after Stalin's death in 1953. With the satellites, he offered a muscular display of his imagination, leadership, and judgment.

And there was more. On November 3, after Laika flew, the USSR rolled out the TU-114, an enormous four-engine, turboprop airplane capable of carrying more passengers farther than any other in the world. Newspapers around the world published photos of the plane that showed it looming over the figures of two men. Three days later the USSR would unveil the world's largest helicopter, an enormous metal grasshopper called the Mi-6, which could lift three times more weight than any copter made in the West.

In an era when brutish hardware ruled the conventional battlefield, the USSR demonstrated that it possessed the biggest and the best. Granted, it had pulled everything off with a little bit of luck. From the first successful flight of the world's first ICBM in August through two *Sputnik*s and the demonstrations of the giant aircraft, no crashes or launch pad explosions had marred the campaign. It was a remarkable string of achievements.

Khrushchev was flush with the excitement of his country's success on the night of November 6 when he spoke inside the Kremlin's Grand Hall to open the celebration of the anniversary of the revolution. He playfully declared, "Now our first Sputnik is not lonely in its space travels." He then invited his Cold War adversaries, and the rest of the world, to join the crowd in orbit. "Our sputniks are circling the world and are waiting for American and other sputniks to appear and make a commonwealth of sputniks," said Khrushchev. Unable to resist a little poke at his adversaries, he also showed that he knew America's Vanguard was in trouble. "Life has shown, it was the Soviet sputniks," he quipped, "which were ahead in the van."

The next day Khrushchev spent hours on a reviewing stand, flanked by Mao Tse-tung, who had accepted an invitation to visit, watching the massive parade of military units and civilian workers that flowed through Red Square. Mao was so impressed by this display, as well as the *Sputniks*, that before leaving for home, he would declare the Soviet Union to be the "head" of world communism.

In the evening after the parade, in the same palaces where tsars had once danced, vodka flowed freely at celebrations of the state. Chairman Khrushchev, considered perhaps the hardest-drinking ruler in the world, was in his glory. He obliged a starstruck woman with a whirl on the dance floor. He toasted the revolution. And he led the hall in song. The raucous celebration was broken by just a single formal speech, made by a lower-level party official, who honored the brilliance of Soviet space scientists.

While they were delivering praise, the leaders of the USSR should also have given themselves credit for managing the public relations aspects of the race into space. Throughout the Cold War, imagery would be nearly as important as events, and in this case the Soviets handled things almost flawlessly. Khrushchev had administered one psychological blow after another while the politicians and bureaucracy in the United States offered little response.

No Madison Avenue consultant could have designed a more effective campaign. First the Soviets cleverly forecast each step without tying themselves to a schedule. Second, they made sure that the world could hear their satellites on the radio, thereby proving their achievement. Finally, in the case of *Sputnik II*, their choice of a cute little dog instead of a rat or a guinea pig to be first in space was a bit of PR genius.

From Argus in Homer's *Odyssey* to FDR's Scottie Fala, dogs had earned their place in the human heart, and the smaller and less-threatening the dog, the more likely the world would respond with delight to the news of her achievement. A week after Laika's blastoff, writer Gay Talese added Laika to the ranks of the dogs of history in *The New York Times*. He noted Argus and Fala, as well as Vice President Richard Nixon's cocker spaniel Checkers, made famous five years earlier when candidate Nixon admitted he was a gift during a speech he gave to refute charges that he had accepted an illegal campaign contribution. Of Laika, Talese wrote, "Man's best friend has stolen the headlines." This was true, even in Huntsville, Alabama, home of the Army missile program, where the local paper paid tribute to Laika in its Sunday "Dog of the Week" column. On November 11 *The Huntsville Times* feature, which always included a photo of a cute canine available at the local pound, showed a puppy seemingly suspended from a parachute. According to the caption, he was a refugee from the Soviet space program.[2]

Of course, with the affection people reflexively grant to dogs, came concern for Laika's welfare. Here the Soviets suffered the first setback on their space propaganda campaign. Early reports stressed Laika's healthy condition, describing her as "calm" and in "generally satisfactory condition." In America, experts debated the possibility that Laika might return to Earth safely. Vanguard program scientists discussed the techniques required and theorized that a rocket engine might be used as a brake to slow reentry. But Wernher von Braun's colleague Willy Ley dismissed this prospect, saying he was almost certain the Soviets were not capable, adding that even if they could get Laika safely to ground, locating the capsule once it landed would be extremely difficult.

With expert opinion quickly coalescing around the notion that Laika was doomed, national humane societies in America, Great Britain, and many other countries lodged formal protests with the Soviet government. They were joined by many angry pro-animal activists, including Mary Riddell, president of the Bide-a-Wee Home Association of Manhattan. She was among the first to note the obvious, that returning Laika to terra firma was impossible. "Your Government," she wrote in a letter to the Soviet embassy, "once again proved its inhumanity."

Things were actually much worse than Mrs. Riddell imagined. The truth of the matter, which wouldn't be revealed for decades, was that Laika probably died from heat exhaustion, and perhaps stress, within hours of beginning her mission. As *Sputnik II* soared over the Soviet Union for the fourth time, the instruments that checked her vital signs showed Laika had died as both the temperature and humidity in the capsule had steadily increased. (It turned out that the Chief Designer's team had failed to create an adequate cooling system for the capsule.)

Many days would pass before Laika's demise would be reported officially. In the meantime Laika's inevitable fate was, for most people, obscured by the playful spectacle of a dog in space. Photos and drawings of the dog appeared in the press for weeks. In the Riviera town of Rapallo, Italy, officials announced they would erect a statue in Laika's honor. In the Soviet Union, packs of Laika cigarettes went on sale to commemorate her achievement.

American officials said nothing about the ethical implications of killing a dog in space. This silence was probably due, in part, to the fact that the United States itself was relying on a menagerie of animal test subjects in its own space research. As Laika was circling the globe, the U.S. Air Force was settling four black bears from Catskill, New York, into their new home at Holloman Air Force Base in New Mexico. Officials first denied the bears would be used in high-speed sled tests. A week later a bear named Oscar was knocked out with anesthetics, strapped into the chair of a high-speed sled, and sent on a ride that subjected him to twenty times the force of gravity. "We wanted to prove that a person could withstand rapid deceleration with no ill effects," said a military spokesman. A thorough examination being necessary to prove the absence of ill effects, Oscar was killed and then autopsied.[3]

★

In the case of *Sputnik II*, it wasn't the dog that impressed President Eisenhower; it was the weight of the object the Soviets had put in orbit.

Shortly after *Sputnik II* was announced, and confirmed by U.S. observers, Herbert York saw the president in a hallway outside the Oval Office. Ike asked York, then chief scientist at the Pentagon's Advanced Research Projects Agency, to estimate the mass that was now circling the United States every few hours.

"I'm certain it weighs over a thousand pounds," said the scientist.

"How do you know that, York?"

"The Russians said that's what it weighs and I believe them."

"You're the first person who ever told me he believes what they say."

"Well sir, the Soviet academy said it. Usually they don't say anything about these things, but when they do speak, they tell the truth."

Nearly fifty years later, York could recall exactly the words he exchanged that day with Eisenhower, and the president's mood. Ike was impressed by the Soviet achievement, a bit frustrated with his own rocket programs, but not overly alarmed about the military implications of the second satellite.

As calm as he was about the realities of the USSR's achievements in space, public perception of the president's state of mind was another matter. Eisenhower had endured a month of constant criticism about *Sputnik*, U.S. space projects, and national defense. Many of his critics tried to paint him as not just mistaken, but out of touch. The carping had been accompanied by a flood of advice from old Army associates, business leaders, and various advisers. Some of this advice, like a one-page memorandum sent by Bernard Baruch, who had advised every president since Woodrow Wilson, was actually requested. (Baruch recommended an accelerated or "bulled" program that would prove to the public that "the impossible will be done, if necessary.") But most of it was unsolicited.

The president also had to anticipate the imminent delivery (and leaking to the press) of a troublesome report by a panel on defense preparedness. Headed by H. R. Gaither, the former president of the Ford Foundation, this group was going to make ominous headlines with an alarmist report about Soviet intentions and weapons programs and suggestions for huge increases in American defense spending.

In Eisenhower's own analysis, the Gaither Committee overestimated Soviet potential. Considering secret intelligence not available to the committee, he believed that the USSR was incapable of producing large numbers of long-range bombers. He also thought that the Gaither group had ignored certain American advantages, especially the allied bases that formed a menacing ring around the USSR. But military facts could be drowned out by the sound of so many alarms. And everywhere the president turned, he could hear noise — he would eventually call it "near hysteria" — about the threat America faced and the need for a muscular response.

Among those he regarded as hysterics were Army officials who stoked press reports on the fantastic possibilities that awaited America in space, implying these dreams could come true if the money were available and they were freed of restrictions. In mid-October they promoted as an example Wernher von Braun's idea for a large reconnaissance satellite code-named "Big Brother" that would watch the Earth's surface for signs of a Soviet missile attack and transmit warnings via still photographs or television images.

By the end of the month, the Eisenhower White House had had enough and began to counter both the critics and the rocket schemers more forcefully. On October 29, after the journal *Aviation Week* revealed surveillance of Soviet missile tests and attacked the White House for brushing aside analysts' concerns about Soviet successes, the president sent an aide to confront publisher Donald McGraw. Brigadier General A. J. Goodpaster met with McGraw in the publisher's office. He said the article had contained some inaccuracies, but more importantly, damaged American security. The publisher promised a more careful approach in future issues. (It was no surprise that *Aviation Week* was hyping the Soviet threat. At the time the magazine was practically the house organ of the missile industry, which supplied it with substantial advertising revenues.)

The next day Eisenhower displayed flashes of irritation as he met with Defense Secretary McElroy to map out the first of three "chins-up" speeches he planned to give in the coming weeks. Among other initiatives, McElroy recommended that Ike announce that the Army Jupiter C rocket was being prepared as a backup to Vanguard. Before the meeting ended, a testy Eisenhower reminded his men that he would need a well-reasoned explanation for this policy change.[4]

While the president prepared his campaign of reassurance, Democrats and Army rocketeers kept pressuring him. Two days after Laika flew, von Braun took time out from meetings at the Jet Propulsion Laboratory in Pasadena, where Van Allen's satellite, with its radiation detector and radio, was being prepared, to talk to reporters. Even though Laika was doomed to die, he pumped up the Soviets' achievements—"they have conquered the problems of space medicine"—and he scolded those, like Eisenhower, who weren't alarmed by the *Sputniks*.

"Anybody who is too optimistic today, who says everything is going to be hotsy-totsy, is doing a great disservice to the country," he said. As a veteran of the Nazis' concerted effort to produce the V-2, von Braun hailed the advantages of authoritarian-style rocket science concentrated in "one team, one facility." He also revised history a bit in his own favor, insisting that at the close of World War II America had "no real program" for developing missiles.

(In fact, as von Braun had to know, work on missiles, multistage rockets, and many related technologies was well along in America before the Germans arrived. In the 1940s, JPL pioneers, especially visionary Frank Malina, built rockets that lifted airplanes and the first sounding rocket of the type Van Allen used to explore the upper atmosphere. In 1945 this all-American-made rocket became the first man-made object to reach the edge of space. But while these achievements proved that the Germans had no monopoly on missile development, in the political climate of 1957 no one would challenge von Braun's version of history. Malina, who had attended meetings of communist groups before World War II, had become a victim of Cold War witch hunts. While von Braun and other ex-Nazis were becoming celebrated defenders of America, he had fled to Europe, and faded from public view.)[5]

On the day when von Braun's comments were published, Lyndon Johnson was one of three senators who spent five hours at the Pentagon in discussions with Secretary McElroy and others. When Johnson emerged from those meetings and reporters asked if he heard anything to change his mind about the administration's defense failures, he had a one-word answer—"No." When asked to elaborate, he made it clear that

America had fallen behind in a vital competition and "we have not kept step with the needs of our times."

The very next day, Senator John F. Kennedy used a speech in Topeka, Kansas, to raise the level of criticism. (Kennedy was already preparing to run for president in 1960.) He was on fairly solid ground as he complained about rivalries within the military, but his most cutting rhetoric dealt with style, not substance. "The people of America," he said, "are no longer willing to be lulled by paternalistic reassurances, spoon-fed science fiction predictions, or by pious platitudes of faith and hope."

Kennedy's remarks on space reflected a strategy that had been taking shape among Democrats since the middle of October when a former Senate staffer named Charles Brewton went to Austin, Texas, to meet with a confidant of Lyndon Johnson named George Reedy.

Brewton saw in the space issue a perfect weapon for, as Reedy later said, "clobbering the Republicans." In Austin he had asked Reedy to drive with him into the Hill Country outside the city, where they found a high point that offered hundred-mile views beneath an endless sky.

"He began to talk about the space program; that man had really mastered the drift of it, the poetry," recalled Reedy years later. "He didn't know very much about outer space but he had grasped immediately the fact that this was something that could change the whole way that we lived; it could change our nation. He convinced me. I remember going back that night. My mind was just full of it. I sat up most of the night reading everything that I could. And I wrote the Senator a long memorandum the next day, which went beyond Charley's [Brewton] thinking because I knew a little more about space. In that memorandum, I said that this would go far beyond a mere defense thing. The immediate public reaction would probably be fear, but that long range, this could be one of the great dividing lines in American and world history, the whole history of humanity."

Eventually the Democrats under future presidents Kennedy and Johnson would pursue the historic glories Reedy and Brewton imagined. (Writing in the awestruck style that afflicts so many when the subject of space arises, Brewton had declared at the time that nothing short of a "race for control of the universe has started.") But in the short term, they had chosen to focus on the clobbering. So obvious was the politicking in the statements made by leading Democrats that the peripatetic Scotty Reston of *The New York Times* would soon describe the issues of science and national defense as new touchstones for presidential hopefuls. Back in Washington, after his family adventure in Moscow, Reston wrote that Kennedy, Johnson, and many others, including Senator Hubert Humphrey of Minnesota, were using the issue to gain attention. Among the Democrats, he noted, only Adlai Stevenson, former governor of Illinois and Ike's opponent in 1952 and 1956, offered anything but criticism of past policies.

There wasn't much Eisenhower's political allies could do to help him quiet his critics, especially those who were positioning themselves for the election in 1960. One of the few partisan ideas floated by Republican Party operatives called for shifting attention to the prior Democratic administration. The main point they intended to make was that President Truman's State Department had made it hard for America to get the best Nazis after the war, and therefore the Soviets had scooped them up. Along with being untrue—America's Germans were far better than the USSR's Germans—this argument seemed beside the point and never got much traction. Other efforts, including the series of speeches dubbed "chins-up" talks, would go a little further toward showing that Ike was an engaged and responsive leader.[6]

★

Executives at the ABC television network considered the level of public concern about the *Sputniks*, and stuck with the Walt Disney show *Zorro* for its 8 P.M. broadcast on Thursday, November 7, but NBC preempted Groucho Marx to carry live the president's first "chins-up" address to the nation on the space competition and national defense. Eisenhower appeared at his desk in the Oval Office, where a light-colored rocket nose cone roughly three feet tall and two feet in diameter had been placed on the floor to his left.

In the first half of his talk, Eisenhower sought to reassure the country with a review of American military capabilities that included enough nuclear firepower "to bring near annihilation to the war-making capabilities of any country." In the armaments grab bag were nuclear-powered submarines, an almost entirely new air fleet, and thirty-eight different missile systems either in operation or in development. As Ike pointed out, four battalions equipped with just one of these weapons, the Corporal, could deliver more destruction than all the artillery fired by the entire U.S. military in the Second World War. America had not yet deployed an intercontinental ballistic missile, he admitted, but considering the many medium-range rockets already posted at bases close to the Soviet Union, the nation's need for an ICBM was not pressing.

When he finally turned to the topic of space, the president maintained his old position—the *Sputnik*s weren't a threat—and boosted U.S. achievements. Pointing to the nose cone on the floor, he explained that it had been recovered after a flight several hundred miles high, proving that America had developed materials that would allow objects (like atom bombs) to return from space without burning up. (To the great irritation of folks at the Redstone Arsenal, he didn't mention that the trophy on display in the Oval Office had been fired by the Army and that when recovered it contained a letter addressed to General Medaris that read, "If you get this letter it will be the first letter delivered by missile.")

After his description of America's might, the president then bent to those who demanded the United States match its Cold War adversary. He admitted that science may have been given short shrift by his administration and promised to fix this problem with better management and the appointment of a new science adviser, James Killian, president of the Massachusetts Institute of Technology.[7]

Response to Killian's appointment was universally positive. The stock market answered the president's commitment to space with a rally driven by the shares of companies that made missiles, rocket fuel, electronics, and exotic metals. Political reviews of Ike's speech broke down along party lines. Republicans generally praised it; Democrats said it didn't go far enough. Wayne Hays, a Democratic member of Congress from Ohio, complained about a lack of concrete proposals but added, backhandedly, that "it does sound, though, like Sputnik has shocked the president awake, and maybe he will stay awake."

The administration followed the first "chins-up" talk, and complaints about a lack of specifics, with an announcement that signaled victory for the Army Redstone group over at least some of the "bastards" that General Medaris had damned on the night of *Sputnik I*. Describing it as a "supplement" to Vanguard, Defense Secretary McElroy declared that the Army would pursue a program of six satellite launches with the Redstone-based Jupiter C rocket.

In less than a week, the president and his team had adopted some of the urgency of its critics, demonstrated a new dedication to science and space with Killian's appointment, and capitulated to von Braun and Medaris, giving them a satellite program, even though it violated the president's policy against going to space with military hardware. All of these developments, which signaled Eisenhower's shift from reassurance to action, could be traced to the influence of domestic politics, the tenacity of Medaris and von Braun, and the addition of a second *Sputnik* circling the globe. Indeed, after *Sputnik II* the public's anxiety was far worse than it had been after the first satellite was launched. Polls showed that a surprising percentage—more than a third—believed that existing Soviet rockets were capable of destroying major U.S. cities. This was not true, and would not be for several years, but the facts mattered less than how people felt.

Having finally won, von Braun granted the Associated Press an interview so that he could press his advantage. After predicting more Soviet achievements, including a possible manned spaceflight, he called for big increases in spending for a five-year U.S. program that would pursue his space agenda, including a station in Earth orbit. If anyone had doubts about the justification for all this, von Braun argued that the Soviets were threatening to become "masters of the space around us." And he challenged Eisenhower's vision of peaceful development of outer space. Sounding much like his partner, General Medaris, he said that future military strategy would depend on "space weapons" and "ten or fifteen years from now space superiority will have taken the place of today's air superiority."[8]

★

The president made his second "chins-up" speech on November 13 at the Municipal Auditorium in Oklahoma City. There he told an audience of 6,000, and millions listening to radio and TV broadcasts, that defense spending would be increased to meet the Soviet challenge. He suggested nationwide testing to identify and encourage students with an aptitude for science. And in a sharp display of Cold War hawkishness, he compared Khrushchev and the Soviet Union to the threat posed by Hitler and Germany in the 1930s. "We shall not make that mistake again," he promised.

The president's action-hero words were matched that day by the exploits of the cigar-chomping commander of the Strategic Air Command, General Curtis LeMay. During World War II, LeMay had dealt with daunting German air defenses by flying directly into their fire. After the war, in 1948, LeMay had applied a similarly bold response to the Soviet blockade of Berlin, leading the massive airlift that overcame it. So it was no surprise, really, that he responded to the *Sputnik* challenge by taking to the air again. As the president had traveled to make his speech in Oklahoma, LeMay was setting a distance record flying nonstop from Washington, D.C., to Buenos Aires in an Air Force KC-135. (Since his jet was a tanker that was like an airborne gas station, LeMay had an advantage over the previous record holder, French aviatrix Elizabeth Boselli.) When LeMay returned to America, landing in Washington, D.C., he was met by an Air Force band that played "The Star-Spangled Banner," and by the Air Force chief of staff, who gave him a medal. LeMay said the flight proved the Strategic Air Command's ability to attack the Soviet Union.

The display put LeMay on the front page of *The New York Times*, along with reports on the president's address. But for the casual reader, they both may have been upstaged by a new picture of the space dog Laika in her capsule, front paws crossed, mouth open and ears cutely cocked, that *Times* editors ran at the top of the page. As all newspeople and entertainers know, you can't compete with a dog, even if it is dead.[9]

★

It was about 10:40 P.M. by the time *Columbine III*, a four-prop Lockheed Constellation with its peculiar three-finned tail, zoomed down the runway at Will Rogers Airport to carry the president back to Washington after his Oklahoma City address. He arrived at the White House well after midnight.

For a sixty-seven-year-old president who had suffered a heart attack and endured a major operation for ileitis in the last two years, Ike had been under substantial pressure. Along with the satellite-and-rocket politics sparked by *Sputnik*, he had spent much of the fall in confrontation with Arkansas governor Orval Faubus over the integration of the public schools in Little Rock. After the governor used National Guard troops to bar black students from entering the school, Eisenhower spent weeks negotiating with him to no end. The matter was finally settled when the president sent the Army's 101st Airborne Division to protect the black students, who were finally admitted.

The Little Rock crisis had damaged America's reputation abroad and earned the president new political enemies in the Deep South. It lingered into October, when Washington then became obsessed with *Sputnik*. Although he did escape for regular rounds of golf and to practice his swing on the White House lawn, the president got less than his normal amount of rest during this period. In mid-November, under pressure from his doctor, he left Washington for a few days of rest at a cabin his fellow hackers had built for his use at Augusta National Golf Club in Georgia.

While at Augusta, Eisenhower wrote a long letter to an old friend confessing that he was struggling to maintain his good humor and hold on to "shreds of a once fairly good disposition." Displays of temper were a sign of weakness to Eisenhower, and he clearly regretted when he showed his anger. Given the demands he faced, it would have been more surprising that he ever smiled at all.

"Since July 25 of 1956, when [Egyptian president Gamal Abdel] Nasser announced the nationalization of the Suez [Canal], I cannot remember a day that has not brought its major or minor crisis," he wrote. Some of these crises were dealt with in secret and would remain unknown to the public, he noted. Other problems, like the harsh politicking over space, would have been much easier to handle, he suggested, if he could have mentioned "things that I don't dare allude to publicly."

Ike listed upcoming events—speeches, the State of the Union address, "endless conferences with legislative leaders"—with the weary tone of a young boy discussing his Saturday chores. Then he revealed the source of the main burden of the presidency: his worry about "doing the right thing." Some problems "are so complex and so difficult that there is no satisfactory answer," he added. With others, he wrote, he took comfort in his faith that "the Almighty must have in mind some better fate for this poor old world of ours than to see it largely blown up in a holocaust of nuclear bombs."

After almost a week at his cabin, where he worked as much as he relaxed, the president returned to Washington on a Thursday night. He immediately set to work on the text for his next pep talk for the American people, slated for Tuesday in Cleveland. In a note to Secretary of State John Foster Dulles he asked for ideas to match his theme "waging peace" and asked if it might be a good idea for him to address the same topic at the United Nations.

The peace talk would mark an attempt by the president to shift the post-*Sputnik* conversation away from the space race and militarism and toward ideals that would promote stability. In this way Eisenhower, who was becoming irritated with the constant public comments of generals, admirals, and scientists, might gain control of the nation's attitude toward space exploration and security.

On the day before he was supposed to speak in Cleveland, the president rode in an eleven-car motorcade across the Potomac to National Airport to greet the king of Morocco, who was coming for a state visit. Eisenhower sat in the back of one of the White House limousines, which were called bubbletops because they featured clear plastic bubble roofs. Although it was cloudy, windy, and cold (mid-forties), the bubble had been removed, exposing Ike to the open air.

At the airport, a fidgety president stood on the apron of the runway, impatiently shifting from foot to foot as he waited for the king's plane, borrowed from the U.S. Air Force, to arrive. Ten minutes later, the plane touched down and then taxied to where the president and motorcade waited. Stairs were rolled up to the door of the airplane. Dressed in a blue-gray robe and white sheepskin slippers, the king descended to the tarmac and the two leaders stood together in the breeze for two national anthems, a twenty-one-gun salute, and several brief speeches. Eisenhower then rode with the king in the open limo, which took them to Blair House, where his guest would stay. By the time he got back to the

White House the president didn't feel well—he believed he had caught a "chill"—and went to his residence to rest.

Sometime before 3 P.M., as he returned to the Oval Office and sat down to work, the president suddenly felt dizzy and found he was unable to pick the top sheet of paper from a pile on his desk. When he finally managed this and started to read, the words on the page looked like they were spilling off the paper. Two or three tries at picking up a pen failed, and then, when he stood up, he had to brace himself to avoid falling.

Alarmed, the president summoned his secretary, but when she came he was only able to speak a garbled mess of words. Arriving soon after the secretary, Eisenhower's aide, General A. J. Goodpaster, recognized something serious was going on and said, "Mr. President, I think we should get you to bed."

Goodpaster took Ike by the arm and helped him to the residence and into bed. The president's personal physician, a seventy-six-year-old retired general named Howard Snyder, hurried to the bedroom and quickly examined his patient, who was remarkably relaxed and unalarmed. He gave Eisenhower a sedative, and Ike slept until the early evening. When he awoke, he was much improved, his speech returning along with his coordination. He alarmed his wife and doctors, however, when he got out of bed by himself, donned a robe, and began shuffling around. When he announced he intended to go to the state dinner planned for the king of Morocco, he was quickly overruled by his wife and physicians.

That night Vice President Richard Nixon accompanied Mrs. Eisenhower to the state dinner. White House reporters clamored for details on the commander in chief's condition. Although five doctors had attended him by this time, and they knew his symptoms suggested not a chill but a minor stroke, no mention was made of this event. Instead the questions raised by journalists were met by evasions and silence.

The truth would come out the next day, as the president's chief of staff, Sherman Adams, and Vice President Nixon assumed some of Eisenhower's duties while insisting he was minimally affected by his illness. They said that although he had trouble finding the right words when he spoke, the president was strong, alert, and still able to communicate. Nevertheless, rumors that the president was more seriously ill and might even have to resign circulated in Washington. These concerns were eased when, after three days' rest, the president had the king in for a visit and signed some official papers to signal he was recovering. By the end of the month he would be well enough to travel to his Gettysburg farm, where he switched on the television to watch the Army and the Navy compete in their annual football grudge match. Thanks to a trick play recommended by chief scout Steve Belichick, this interservice battle was won by the Navy.[10]

FIVE

Hold It!

For more than a month James Van Allen had struggled to stay in the satellite game in a long-distance way, but security rules limited the messages he could send and receive while aboard the *Glacier*. Between rockoon launches, which resumed when he recovered from his infection, Van Allen had plenty of time to worry about being eclipsed by other ambitious space scientists who might be trying to attach their projects to various military missile programs. He was relieved at the end of October to get a telex from the Jet Propulsion Laboratory in Pasadena, which asked:

WOULD YOU APPROVE TRANSFER OF YOUR EXPERIMENT TO US
WITH TWO COPIES IN SPRING?

If he was reading it right, the note from JPL director William Pickering meant that Van Allen's cosmic ray gear would ride inside the first three satellites that the Army team would attempt to launch. On November 10, when he finally reached Lyttelton, the port at Christchurch, New Zealand, Van Allen was able to confirm this news with his students in Iowa. They told him that his cosmic ray experiments were already being prepared for the Army launches. Pleased with this turn of events, he began the long journey that would bring him home for Thanksgiving.

In an ideal world, Van Allen might have preferred for the Navy team and its Vanguard to win the intramural rocket race. He thought the current political uproar could be quieted by an immediate American success, even if it did involve nothing more than a radio beacon in orbit. And given General Medaris's arrogance—Van Allen considered him an especially difficult, "crusty old guy"—this outcome would have fit the professor's Midwestern sensibility, which favored humility and collegiality over ego-fueled aggression.

But for the moment at least, Medaris's group possessed the better rocket for launching Van Allen's equipment. "The difference was von Braun's drive, and greater resources," explained Van Allen decades later. Although the general public wouldn't have known it, von Braun was not much of a rocket scientist, added Van Allen, but he was a skillful leader of a gifted engineering group. "His buddy Ernst Stuhlinger [who held a Ph.D. in physics] was the only real scientist," added Van Allen, "but as a group they knew how to design things and get them built."

From the first contact he had with him at the White Sands Proving Ground in the late 1940s, Van Allen was impressed by von Braun's powerful personality. The German's ambition, which had gotten him charged with sabotage for pursuing his own agenda under the Nazis, had been given freer rein in the United States, where it was far easier to disobey orders and get away with it. Ever since the Pentagon had told his group to stop satellite launching work, von Braun had, with Medaris's help, defied Washington and proceeded with it anyway. Van Allen's Iowa group had been drawn into this work in early 1957, welcoming emissaries from Alabama to the Iowa campus and traveling to the Redstone Arsenal to consult on designs. Months before the first *Sputnik* roared into space, they had already settled all the important details.

The early decision to hedge his bets because, as he explained, "Vanguard was already in deep trouble," suggests that Van Allen was equal to the games that the ambitious played if they were going to succeed in the ferocious competition for space resources. For more than a decade Van Allen had worked to position himself, a physics professor from a state university in Iowa, to get his instruments aboard the first American rocket capable of putting them into orbit. If this required him to work with the Navy and the Army at the same time, he would do it.

While work was done on the cosmic ray instruments, the Army's Redstone group was moving as quickly as it could to prepare its Missile 29 for the JPL/Van Allen satellite. The job got much easier when Washington ended the restrictions on overtime that had been imposed as part of the president's balanced budget policy. As engineers, officers, and technicians found themselves more fully engaged in a campaign to catch the Soviets, morale at Redstone improved dramatically.

While their team readied the rocket, Medaris and von Braun battled with their superiors in Washington. They wanted a guarantee that even if Vanguard succeeded in getting to space first, their rocket was going to be used in an attempt to put America's first real scientific payload into orbit. Medaris later confessed that he "blew up" during heated negotiations over this point. At the lowest moment, both men threatened to resign if they weren't guaranteed at least one certain shot at going into space.

Medaris and von Braun reached this point of exasperation after more than a full year of being constrained from using a rocket that was waiting to fulfill its purpose. Their mood wasn't helped when, after *Sputnik I*, Soviet scientists said that *even they* couldn't understand why the great von Braun and his machine had been grounded. Von Braun's colleague Ernst Stuhlinger heard about the Soviet view when he attended an international conference and chatted with Leonid Sedov, who had worked on *Sputnik*. Had a Redstone been chosen instead of Vanguard, added Sedov, "You would have saved so much time, not to mention troubles and money."

But even if the entire world agreed that they had the best rocket, Medaris and von Braun had to show some restraint. Army Chief of Staff General Lyman Lemnitzer warned Medaris about hyping the race to space while Ike was telling the public no such contest was being waged.

"The time for talking has stopped," said Lemnitzer, as he ordered Medaris to keep his people out of the press. Medaris promised to do his best, but cagily reminded Lemnitzer that he couldn't control civilians, except when it came to official secrets. More days of haggling followed, but before the month of November was out, Medaris would get what he wanted: firm backing for a satellite launch on, or about, January 29, 1958. This date was not made public.

The weeks to come would be the most frenzied in Medaris's career. Along with prepping Missile 29 for a satellite, he was supposed to get two more rockets ready for a top secret mission. These would be sent to the South Pacific, where they would carry atomic bombs into the atmosphere to test Nicholas Christofilos's electron shield hypothesis.[1]

In the meantime, the entire rocket and satellite field would get a boost as Lyndon Johnson's Preparedness Investigating Subcommittee finally opened hearings intended to dissect the failings of the Eisenhower administration, plan a response, and rally the nation behind its scientists and engineers. Over time, the committee would explore supposed failings ranging from education policy (not enough science in the classroom) to moral values. But the immediate task was to sound an alarm about the communist menace.

★

"It would appear that we have slipped dangerously behind the Soviet Union in some very important fields," announced Johnson after the hearing was called to order in a room in the Senate Office Building. *Sputnik* was a more urgent historical event than Pearl Harbor, added the senator from Texas. "But to me, and to every other American, a lost battle is not a defeat. It is instead a challenge, a call for Americans to respond with the very best that is within them."

The committee's first witness, the father of the hydrogen bomb, Edward Teller, had started the week with his face on the cover of *Time* magazine. Inside, *Time* presented him as uniquely wise about science and geopolitics. Pictured with his family gathered round a chessboard—certifying their egghead status—Teller was described as so dogged by Cold War fears he was unable to pursue either pure science or happiness. "The Russians can conquer us without fighting," he warned in the article, if only they emphasize science and technology.

At Johnson's hearing, Teller spoke with his usual pessimism about the need for fallout shelters, stockpiles of food, and individual radiation detectors for the citizenry. Teller darkly predicted a decade of Soviet superiority in science, and not surprisingly, urged steps to raise the status and public regard for American science and its practitioners.

Teller's testimony opened a process that promised to continue for months as Johnson planned an extended investigation that would allow him to hear from more than seventy witnesses, and would keep him and the politically charged matter of missiles, satellites, and defense in the public eye. By beginning this long-running engagement with the fiercest hawk among the nation's atomic scientists, Johnson guaranteed that the next day's newspapers would carry front-page stories about a Democratic Congress addressing a great peril.

Whether Teller possessed real bona fides on Cold War political and military strategy, as opposed to his field of expertise (bomb making), was not discussed. By the mid-1950s scientists had become oracles of public policy as well as technology. This was true for doctors fighting polio as well as physicists splitting atoms. All were courted by the press as well as politicians. Even before he addressed the Senate subcommittee, Teller went on the CBS television program *See It Now* to tell a much larger audience that America was in a "sad and dangerous" situation.[2]

In general, the attitudes of America's more prominent scientists were so well known that TV producers and senators alike could reliably call on certain ones to get the kind of comment they desired. If instead of Teller he had invited Eisenhower's new science adviser, James Killian, to speak, Lyndon Johnson might well have heard what Killian later said he believed, that the country was overreacting to *Sputnik* because it had the Cold War "vapors." This condition caused people to forget that American science was vastly superior to the Soviet Union's. Indeed, in the competition for Nobel Prizes in science since 1930, the United States led the USSR by a score of thirty to one. The vapors also seemed to make people accept claims that the USSR graduated more engineers annually than America, without considering that the quality of their training fell well below American standards. You could call anyone an engineer. Whether he or she could perform like one was an entirely different matter.

Ultimately the Johnson committee would sit for a total of fifteen days in November, December, and January. But the first round was kept short. Just three days were allowed, enough to give the nation a dose of high anxiety. Not surprisingly, during those three days the head of Vanguard, who brought a model rocket to the hearing room to use as a visual aid, said he could have done better with more power and more money. An Air Force general said that interservice rivalries, like the one between the Army Jupiter C group and Navy Vanguard, had exceeded "healthy competition." (In fact, Army missile program leaders spent a great deal of time exchanging letters marked "secret" that dealt not with surpassing the Soviets but with publicity concerns and beating out the Navy and Air Force.)

To counter their critics, administration witnesses who spoke to the committee admitted being surprised by the *Sputnik*s but then announced they would deploy medium-range, nuclear-tipped missiles in Europe within the year. These Jupiters were not the most advanced U.S. rockets and would soon be deemed obsolete, but they were symbolically important. Then, on November 27, the hearings were suspended. For the next ten days, the attention of everyone who was concerned with missiles and satellites and the Cold War would shift to Cape Canaveral.[3]

★

After the Eisenhower administration announced in November that the next scheduled Vanguard test would instead be a full-fledged satellite shot, news editors and reporters across the country began planning to cover it. Other than warfare or a landmark murder trial, it would be hard to imagine an event with more dramatic potential. TV networks organized crews complete with cameras and celebrity correspondents, and major newspapers arranged for star reporters to camp out for the story.

Three hundred miles north of the Cape, in Albany, Georgia, a twenty-three-year-old announcer at a tiny TV station called WALB, where Saturday wrestling matches were broadcast live from the station parking lot, had decided he would cover the big story too. As a ballroom champion (tango and samba), Jay Barbree was actually a much more accomplished dancer than reporter. And he worried that his syrupy Southern voice was a handicap. But he was profoundly ambitious, and eager for an adventure that would take him out of Albany and into the thick of a big event.

On *Sputnik* night, Barbree had spent time with his friend Gene McCall, a college student who had taught himself enough about broadcast technology to become an engineer at the station. Barbree and McCall were dating sisters and these young women joined them for an evening date that involved a parked car and, this time, some actual stargazing as the four of them looked for signs of the satellite.

"I wasn't afraid or really surprised by *Sputnik*," recalled Barbree, almost fifty years later. "What I felt was the excitement, and I knew that this was the beginning of a story that was going to be important for a long time to come."

With no backing from his bosses other than a promise to air his reports, Barbree drove his 1954 Chevrolet coupe for more than eight hours on back roads and two-lane highways to get to Cocoa Beach before the Vanguard launch. He rented a room at the Starlite Motel and struggled to acquire the language of the space community and get up to speed on the particulars about various rockets—Vanguard, Jupiter, Thor, Navaho—and to learn to read the signals that helped local reporters know what was happening.

Barbree heard about the red warning lights that were lit near the pads and the ball hoisted on a staff to warn fishermen about an impending launch. But as an out-of-towner who was barely a reporter, he would need time to fit into the local press corps, which included veterans like the wire service correspondent Vern Haugland of the Associated Press and small-town go-getters like a nineteen-year- old radio reporter named Wickham "Wickie" Stivers.

In a time when the noted financial reporter Sylvia Porter was praised for being "vivacious and eye-filling," Stivers matched the archetype of the beautiful and smart gal reporter. Blue-eyed and petite, Wickie was poised and charming in a distinctly Southern way. (Given that her great-grandfather had been a mayor of New York, her social gifts may have run in the family.) Male competitors who were distracted by Wickie's looks could underestimate her ability. Her daily half-hour broadcast of news and interviews on a local radio station gave her a wealth of local sources, who tipped her off to developments at the Cape. She could also count on friends at the Patrick Air Force Base officers club, whom she met through her boyfriend, a military dentist. The club occupied a spot on the beach that made it a convenient stop on her way to and from work. It was open to civilians only by invitation. Thanks to the dentist, a commissioned officer, Wickie was always welcome.[4]

Unfortunately even the best connections wouldn't help Wickie or anyone else get much information about the big Vanguard launch. Already irritated by previous accounts of their "ill-fated" quest for space, the rocket group protected itself from scrutiny behind the Air Force public affairs office, which offered few clues about what was going on. (Because it was in charge of operations at the Cape, the Air Force handled public affairs for all launches.) This left the growing horde of news reporters little option but to besiege a recently arrived public affairs officer named Kenneth Grine.

Before coming to the Cape area, Grine had had little direct involvement with space projects. His closest contact came while he had worked for a classified military journal and received the unusual assignment of researching names the Air Force might use for its missiles. It was Grine who went to a public library in Baltimore and settled on the gods of mythology—Thor, Atlas, and Titan. At the time he didn't talk about this quirky achievement. Nor did he discuss his recent surgery for a brain tumor, which had left him with a serious reading disability. It was a devastating loss for a man who loved books and worked with journalists. Somehow he managed to accommodate his disability without anyone catching on.

In general, Grine didn't receive much reading material from the research and development programs he was supposed to cover. This was done intentionally by military officers and civilian supervisors who wanted to keep their secrets. And Grine saw very little actual missile-related activity because he was stuck in an office at Patrick Air Force Base, roughly twenty miles from Cape Canaveral.

"All the newsmen wanted information and they were all over our offices," recalled Grine long afterward. "But I didn't even have a pass to go to the Cape. The Department of Defense had put all this pressure on my boss, General [Donald] Yates, to keep things controlled, and he did. Almost everything was classified."[5]

For example, when a late-November night launch of a Jupiter rocket ended with a fiery explosion in the sky, the Defense Department would report only that "the missile failed to complete its programmed flight because of technical difficulties." Grine, who recalled that his phone never stopped ringing in this time, could offer even less information on two other launches that day. One was called simply a "test vehicle." The other was not identified in any way.[6]

Often the military didn't even announce a launch was planned. In other cases, a general notice about a launch—say, one scheduled for a particular morning—would be followed by hours of unexplained delay. Reporters had to resign themselves to long periods of sky watching. Although a lucky few found space in nearby motels or rented cottages with good sight lines, in this time the best option for any curious observer, be he journalist, local citizen, or tourist, was to occupy a stretch of public beach that overlooked the launch pad area to the north and wait. At this spot, which some called Bird Watch Hill, photographers guarded cameras on tripods while reporters scanned the sky for smoke and flame. Regulars like Wickie Stivers knew where to fix their gaze and developed a sense of how things worked that allowed them to rest or play cards while keeping one eye on the horizon. A newcomer would invariably be told to focus on the most visible missile-shaped object on the horizon only to learn, after the launch, that he had been gazing at a striped lighthouse.

If a launch was delayed, the rocket watch could continue past sundown. Reporters who were required to keep editors informed would make periodic walks to the nearest pay phone and hope that nothing happened while they were gone. When a launch was canceled, the ball would be pulled down from its mast at the port and the red lights near the pads would be turned off.

Whether observers actually saw something fly on a given day, or simply logged empty hours on the beach, relief waited in the bars, restaurants, and motels along Route A1A. Once every week a big poker game brought those who were involved in the missile projects (contractors and even a few military officers) together with reporters at a two-story motel suite occupied by the Associated Press, which had a balcony overlooking the beach. On other nights the insiders and outsiders drank and danced together in the lounges at the Starlite and Vanguard motels. Both lounges often offered live entertainment to draw crowds. Nightclub acts were

given reviews in the national media—Tammy Grimes was big in 1957—
and performers could be tapped for Broadway and film.[7]

Of course, not every resident of boomtown was up to the demands
of nightlife on the motel strip. Some were a little too old. Others were a
little too married with children. And even the regulars at the Starlite
could opt for a night in. But it could be hard to escape missiles, satellites, and Cold War fear, which were regular topics on news programs
broadcast by the national networks. Typical was Mike Wallace's televised interview of recently retired defense secretary Charles E. Wilson, a
budget-cutter who had been criticized for neglecting science in general
and Vanguard in particular.

In an exchange typical of his style, Wallace declared the Soviet military to be superior to America's and bluntly accused Wilson of slighting
research. Wilson challenged the whole idea that the USSR had better
forces and argued that the Soviets' main achievement had been manipulating the American public.

> *That's the thing about our people. They're so cracked loose in
> the Buck Rogers age that they're seeing things. Even the Texans
> they're seeing space ships and flying saucers . . . it's what the Rus-
> sians thought would happen. They intrigue me. They understand
> us so well.*[8]

★

The Soviet propaganda had its most direct effect on the morale of the
members of Vanguard team, who were stuck with the buck as it was
passed in the weeks after *Sputnik*. In October and November they labored furiously to get America into space for the first time and reclaim
some prestige for the West. Proof of this effort came in mid-November
when the tall slender rocket called TV-3 (Test Vehicle-3) was wheeled
out of its hangar, erected on its stand at Launch Complex 18, and teth-
ered to a red and white gantry tower. Most of the rocket was either me-
tallic gray or black, but a band of white encircled its middle. Long strips
of black rubber dangled from its flanks. These "spoilers" would protect it
from the von Karman effect, a potentially disastrous, destabilizing phenomenon caused by air flowing over a rocket fuselage. The von Karman

effect was so difficult to understand that few in von Karman's field could explain it. It was the product of a man who was revered for his brilliance and fulfilled every definition of the eccentric and absent-minded scientist. He liked to wear capes, gestured wildly during lectures, and, when concentrating, often chewed on his handkerchief. During World War II, he was accompanied on official trips by an aide whose main job was collecting secret papers von Karman left in taxis and hotels.

The countdown to firing commenced before dawn on December 4, a day that began overcast but promised clearing by the time the rocket was to fly. More than a hundred reporters and photographers from around the world gathered hours early on the beach, just in case a prelaunch problem led to an explosion. Other gawkers joined them in a steady stream until there were, according to one account, thousands of people on the sand, patiently waiting without any sense of how things were going at the launch site.

The checklist of tasks that would have to be accomplished in precise order prior to ignition was more than twenty pages long, and many of them were either dangerous or complicated, or both. Highly volatile fuels had to be pumped from tanker trucks into the rocket. Leaks were common, and the liquid oxygen, LOX, which was maintained at lower than minus 270 degrees Fahrenheit, wreaked havoc with valves and pipes. On this day, after a series of minor problems, the LOX froze a valve so completely that the fuel spilled from the piping and created big clouds of vapor and blocks of ice on the test stand.

The struggle with the LOX system and rocket went on all day. Workers who had reported for duty shortly after midnight staved off exhaustion with coffee and hamburgers bought at the big white lunch trucks known as "garbage trucks" that drew huge crowds every time they pulled up to the site. As night turned to day and then night again, workers who weren't actually performing some task rested on the concrete outside the blockhouse or napped on benches.

Inside the blockhouse, engineers and technicians sat in front of large consoles—steel cabinets set on tabletops—and watched dials that gave them readings on the condition of the rocket's various systems. The denizens of the blockhouse were, to a person, white and male. Those who were seated wore bulky headsets that made them look a little like telephone operators. Behind them stood other members of the team, who stared over shoulders and glanced from console to console. Cramped, lit by artificial lights, the room had a submarine feel—hot, stuffy, claustrophobic.

As the crowded space filled with tobacco smoke, tension, and nervous chatter, the Navy's liaison fielded so many calls from impatient higher-ups that he often held two phones at once while a third was kept on hold. Gradually it became obvious to everyone that no easy fix was possible and fuels would have to be drained from the rocket to allow for a major repair.

"To hell with this," said Vanguard field manager Dan Mazur. "I can't take anymore. No one can take anymore. Let's ship it back to Baltimore."

"Scrub!" shouted launch director Robert Gray. At 10:08 P.M., the countdown was canceled.

"The test has been scrubbed," blared from the loudspeakers outside the blockhouse. "Will the gantry crew return to the gantry and move it back?" Men in jumpsuits and white hardhats scurried to the pad to return the tower to the rocket's side and secure it.

The failure made headlines nationwide. In Iowa City, James Van Allen's local paper, the *Press-Citizen*, reported "Satellite Launching Hits Snags" in large bold letters across the entire width of its front page. The paper also reported that another countdown would begin almost immediately.[9]

In fact, at the time when the *Press-Citizen* was being printed, Vanguard's managers weren't sure when they would be ready for another try. General Yates, who commanded launch support at Cape Canaveral and set the schedule, gave the engineers leeway to begin a countdown as soon as they were ready. At first they thought it would be Sunday, but when repairs to the LOX valve went faster than expected, they set the restart for 1 A.M., Friday, December 6. Word was passed to Washington, where, on Thursday, President Eisenhower attended a National Security Council meeting, then, in the midst of a snowstorm, left for his Gettysburg farm. Although he was making a fast recovery from his stroke, the president planned a four-day weekend of rest.

On the drive north Eisenhower ordered his motorcade to pull over on the road near Frederick, Maryland, to pick up Jerry Beswick, an airman in uniform who was hitchhiking to his base in upstate New York. Ike rolled down the window of the car he rode in to say, "I thought we'd give you a lift." A shocked Beswick hopped into the car behind, which carried the president's physician, General Snyder. Beswick got out at the farm and was taken to catch a bus for the rest of his trip. (For the rest of his life, spent mostly as a husband, father, and car dealer, he would recall the ride as a highlight.) By 5 P.M., after Mrs. Eisenhower arrived by a separate car, the president had settled in for the evening.

On that night, as the president rested and members of the crew reported to the Vanguard blockhouse and launch pad area, millions of Americans gathered to watch *Mars and Beyond*, the third and last installment of the Disney space series, on the ABC television network. The hour-long show offered a sprint through such varied topics as evolution (without controversy), astronomy, and space propulsion. Viewers saw images of a Martian landscape that could have been drawn by Salvador Dali, and interplanetary spacecraft based on Wernher von Braun's designs. He and Ernst Stuhlinger, handsome in matching pale blue shirts, posed with plans for the Mars vehicles.[10]

In the Disney/von Braun universe of the future, a flotilla of spacecraft moved in stately formation through the vacuum of space toward the red planet. Their progress was smooth, graceful, perfect. In the reality of 1957 on the blue planet, the seemingly simple task of firing a rocket was fraught with potential problems. In the first twenty minutes of the second Vanguard countdown, which commenced as scheduled at 1 A.M. Friday, December 6, error readings on control panels sent technicians scrambling to the pad to fix bad electrical connectors. This chore was complicated by the fact that the electrically powered explosives that were part of the rocket's self-destruct system had to be disarmed so the repairs might be done safely. It took nearly five hours to make the repairs. During all this time the countdown clock was kept on hold.

The rising sun was turning the sky pale gray as the countdown resumed. The first birdwatchers gathered on the beach around fires made of driftwood. In the control room, technicians and supervisors worked through the pages of the countdown protocol, anticipating problems that didn't arise. Fuel was loaded without event. Weather reports confirmed that brisk surface winds—sixteen to twenty-two miles per hour—were not too strong to force a hold.

One hour before the launch, deputy launch director J. Paul Walsh picked up a phone in the blockhouse and dialed the White House, where Vanguard's chief administrator, John Hagen, took the call and remained on the line. At about the same time another line was opened between the White House and a hotel room in Gettysburg, where press secretary James Hagerty listened for news. For most of the morning President Eisenhower sat before an easel on the porch at his farm, painting a portrait of his eight-year-old granddaughter, Barbara.

Forty-five minutes before launch, checks of the Cape radar and downrange tracking systems, which John Neilon helped run out of Patrick Air Force Base, turned up no problems. A photographer on hand to document history began shooting film, making sure to get both still and motion picture shots of every control panel and every man in the blockhouse. Outside, the red and white gantry tower was rolled away from Vanguard's side, leaving the rocket standing alone.

Thirty-one minutes before launch, meters showed a sudden spike in voltage in the ignition system. A hold was called so technicians could run out to the pad to find the problem. In half an hour the problem was fixed, the countdown resumed.

Twenty-five minutes before launch, the safety officer closed the big steel doors to the blockhouse. With no more immediate responsibilities, he sat down, pulled a big turkey leg out of his lunch box, and took a bite.

Nineteen minutes before launch, the blockhouse lights were dimmed and nervous crew members were ordered to put out cigarettes and pipes and to stop nonessential conversations. In the half-light stillness, the man assigned to watch a wind gauge called out numbers. The breeze was diminishing.

In the next eighteen minutes the satellite radio beacon, local radar, downrange tracking stations, and the worldwide Minitrack system—including Chester Cunningham's post in Peru—were all checked and found operational. The men at the blockhouse control panels hunched over their meters and dials while the others edged toward the blast-proof window for a glimpse of the rocket. Outside a small stream of vapor puffed from the rocket's flank where excess liquid oxygen was released.

Less than a month before *Sputnik I* shocked the world, Richard Reston, son of *New York Times* writer James "Scotty" Reston, sat behind the wheel of the family's 1957 Ford Fairlane in Moscow. (COURTESY RICHARD RESTON)

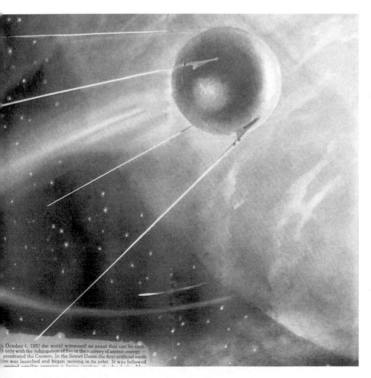

The first man-made object to orbit in space, *Sputnik I* was little more than a ball outfitted with a radio transmitter and swept-back antennae. However, its beeping presence was a triumph for Soviet missile technology. (COURTESY BRADFORD WHIPPLE)

Laika, the Soviet space dog, was the first living creature to orbit the Earth. Her flight put the USSR further ahead of the United States in the space competition. (COURTESY BRADFORD WHIPPLE)

Rushing to catch the Soviets, the United States relied on Vanguard, a mainly civilian project based on Navy technology. But the first Vanguard rocket exploded on the launch pad. (NASA ARCHIVE)

In 1957 physicist Herbert York was the director of the University of California's Lawrence National Laboratory. Chosen by President Eisenhower to head the Defense Department's Advanced Research Projects Agency, he would help lead the American response to Sputnik. (U.S. DEPARTMENT OF ENERGY ARCHIVE)

General J. Bruce Medaris and Wernher von Braun led the group that developed missiles for the U.S. Army in 1957–58. Adversaries during World War II, they became partners in the development of the rocket that would put the first American satellite into orbit. (NASA ARCHIVE)

On January 31, 1958, an Army rocket called Juno I carried the first American satellite, *Explorer I,* into space. *Explorer* contained instruments that helped scientists make significant discoveries about radiation in space. (NASA ARCHIVE)

Left to right: William Pickering, director of the Jet Propulsion Laboratory; James Van Allen of the University of Iowa; and Army missile chief Wernher von Braun hold a model of the first Explorer satellite at a press conference in Washington heralding the successful launch. (NASA ARCHIVE)

The space race drew press to the Cape Canaveral area, and "bird-watching" became a plum assignment for reporters and photographers. Here they are watching the first successful launch of a Vanguard satellite on March 17, 1958. (COURTESY WICKHAM WILSON)

"Girl" reporter Wickham "Wickie" Stivers (*left, wearing shorts*) was one of few women who reported from Cape Canaveral in the early days of the space race. (COURTESY WICKHAM WILSON)

Jay Barbree (*left*) was a young, self-taught radio reporter who gave traffic reports as "Mobile Mike" and moonlighted as a dance instructor before becoming NBC's space reporter in 1958. With him here are Air Force press officer Kenneth Grine and broadcaster Chet Huntley, coanchor of the network's nightly news program. (COURTESY OF JAY BARBREE)

Eighteen-year-old Jo Ann Reisinger wore a futuristic costume to win the Cocoa, Florida, Miss Space contest in 1958. At about this time she began dating reporter Jay Barbree, whom she later married. (COURTESY OF JO ANN BARBREE)

irman Bradford Whipple was a radio specialist at a secret listening post in Germany when he was mong the first to hear the radio beeps of *Sputnik I*. The following summer he traveled with his girl-riend, Anke Hansen, to the 1958 World Exposition in Brussels, where the gleaming Atomium was a entral attraction. There Whipple was recruited by American agents to help steal, photograph, and eturn a model Sputnik from the Soviet exhibit hall. (COURTESY OF BRADFORD WHIPPLE)

er, a squirrel
key like Gordo,
a suborbital mission
ay 1959, surviving
s of force and
ly ten minutes of
htlessness. Unlike
lo, who perished on
ight, Baker was
vered and lived
1984 at Redstone
nal, where she was
exhibited for
rs. (NASA ARCHIVE)

Soviet propaganda at the Brussels was filled with space imagery. Th were toy Sputniks, and Laika the appeared on Soviet cigarette pack (COURTESY BRADFORD WHIPPLE)

A leader of the mission control crew, Curtis Johnston was known for wearing a "lucky" cowboy outfit on launch days. Here he poses at Cape Canaveral with a hobby horse he received as a gift from colleagues. (COURTESY CURTIS JOHNSTON)

Five minutes before the launch, the pad area was cleared of workers. The voice of the countdown announcer echoed through loudspeakers outside. Inside, competing voices noted that the fuel pressure, maintained by complicated plumbing, was good.

"Hold!"

At the one-minute mark the test conductor stopped the count for another weather check. A cycle of stiffer breezes had begun. After a few minutes the wind decreased. The count resumed, called by an announcer, second by second.

In the carefully choreographed minute that followed, microphones transmitted to the blockhouse the sounds of the lines that supplied the rocket being disconnected and falling away. A hose that delivered helium dropped off and fell into a net positioned to catch it. Another that sent cold air to the satellite, to protect it from the heating rays of the Florida sun, also fell. The valve that let liquid oxygen vapor escape from the rocket was shut. The white vapor that looked like steam spewing from a kettle quickly dissipated.

With ten seconds to go, engineer Kurt Stehling noticed the men near him were sweating and their body odor, already intense, grew worse. He saw that that Paul Karpiscak, charged with actually firing the rocket, had poised his finger over the toggle switch.

"Stop your goddamn lights blinking and flashing," shouted Dan Mazur as the photographer recorded the moment. "The photos can wait."

Five seconds from ignition, out on the pad, water sprayed to clear fumes made a rainbow near the rocket.

"T-minus four seconds, T-minus three seconds, T-minus two seconds, T-minus one second."

"Fire!"

At the instant when Karpiscak hit the switch, automatic systems took over. Oxygen and kerosene held in Vanguard's tanks were sent to the first-stage engine. Fire and smoke billowed from the base of the rocket. From his post in the blockhouse, the assistant launch director relayed the news to John Hagen at the White House, saying, "Zero, fire, ignition!" Then, as eyewitness Stehling would describe it a few years later:

The engine started with a heart-rending, hoarse, whining moan like that of some antediluvian beast in birth pain. Flame filled the nozzle at first and then built up with a great white crescendo to a tremendous howl, brilliantly white, streaked with black. The vehicle shook itself momentarily, like a wet dog. Ice and snow fell off the sides. The banshee howl of the engine increased. The vehicle hesitatingly lifted itself loose from its iron womb and rose slowly. We rose with it, on our tiptoes.

"*Look out! Oh God, no!*"[11]

★

On Bird Watch Hill, where hundreds of reporters were joined by many more citizen rocket-gazers, the smoke from the rocket engine would be visible before the sound carried across the water to the shore. Camera-operators stared through telephoto lenses. Reporters equipped with binoculars raised them to their eyes. Fortunately for them, the Vanguard pad was at Launch Complex 18, which was in the cluster closest to the south boundary of the missile range and quite visible.

The big challenge for the reporters would be access to phones, so they could get the news out once something happened. This was especially true for broadcasters, who were enormously competitive and determined to be first on the air with the news. Vanguard marked the first time that major networks invested substantial amounts of money and manpower in covering a space shot live. Within a few years they would be spending millions of dollars to cover each big launch. But this time they made do with small crews built around lone correspondents.

The young CBS reporter, Harry Reasoner, and his producer, Charles Von Fremd, had gained an advantage in the race to be first on the air by renting a bit of real estate that had become extremely prime: a cottage with a view of the Vanguard pad. While cameraman Paul Rubenstein took his post on the beach, where the sightlines were best, Reasoner stood on the porch at the cottage and peered through binoculars. Inside the little house, Von Fremd's wife, Virginia, held a phone that was connected to the network newsroom in New York. Reasoner never forgot what happened at 11:45:59 that morning:

I saw an unmistakable flash of flame and the pencil-thin white rocket began to move.

"There she goes!" I shouted.

"There she goes!" shouted Virginia into the phone.

"There she goes!" shouted the executive in New York, hanging up the phone and charging off to get the bulletin on the air.

We beat ABC and NBC certainly. There was only one problem.

A tenth of a second after I shouted, "There she goes!" I shouted, "Hold it!"[12]

Cub reporter Jay Barbree, tall, lanky, and curly-haired, had stationed himself closer to the launch site, at a bank of six pay phones that stood at the gated entrance to the missile range. With two newspapermen, who didn't need instant access to their editors, he shared the duty of holding the phone all morning long, so that it was available at the key moment. From that spot he could see only part of Vanguard TV-3 in the distance. Having never witnessed a rocket launch, he wasn't entirely certain about what he was seeing. Nevertheless he was filled with excitement.

"I was on the phone live with the station back in Albany, Georgia," recalled Barbree. "I know I started out saying, 'There's ignition! We can see it!'"

After the delays in midweek, the mere fact that the rocket was fired brought cheers to the gate area as well as Bird Watch Hill. More cheerleader than reporter (the launch occurred just before he took a journalism correspondence course), Barbree got very excited. But then, in an instant, the spell was broken.

" 'There's no liftoff,' I told them. I said, 'It's failed. Vanguard is breaking apart. It failed, ladies and gentlemen, Vanguard failed.'"

Generations later, stories would be told about radio and TV reporters who mistakenly told listeners that Vanguard had risen with such explosive speed that it became invisible to the human eye as it streaked into space. No solid evidence of this kind of misstatement seems to exist. Instead, those who watched confessed that they were fooled for a moment and then felt stomach-churning disappointment and fear for the launch crew as the rocket wobbled, its nose cone shifted, and then the body of the rocket collapsed in on itself, like a city building detonated by expertly placed dynamite charges.[13]

Vanguard TV-3 actually rose a few inches, perhaps a few feet, off the pad before it collapsed back on itself. Inside the blockhouse J. Paul Walsh cried "Explosion!" into the telephone receiver and in Washington, John Hagen, on the other end of the call, replied "Nuts!" He immediately relayed the news of Vanguard's demise to press secretary Hagerty in Pennsylvania.

Given the smoke and dust that billowed around the base of the rocket as it ignited, the men watching through the thick window in the blockhouse had trouble seeing Vanguard's very short flight. Instead they were dazzled by what Stehling later described as the "brilliant stiletto flames" that flared out in every direction as the roaring sound of the rocket grew.

> *The vehicle agonizingly hesitated for a moment, quivered again, and, in front of our unbelieving shocked eyes began to topple. It also sank like a great flaming sword into its scabbard, down the blast tube. It toppled slowly, breaking apart, hitting part of the test stand and ground with a tremendous roar that could be felt and heard even behind the two foot concrete walls of the blockhouse and the six-inch-thick bulletproof glass. For a moment or two there was complete disbelief. I could see it in the faces, I could feel it in myself. This couldn't be.*

From afar, it looked like the blockhouse was being engulfed by orange flames and a roiling black cloud of greasy smoke. Bits and pieces of rocket—some broken off by explosion, others by the collapse—began hitting the fortress. After a technician pulled the proper lever, an automatic fire control system poured thousands of gallons of water onto the pad. At first much of the water turned to steam, but within a minute the stream was cooling the pad and dousing the flames.

Loosed from the nose of the rocket, the four-pound silvery satellite bounded along the ground. It came to rest, still beeping out its signal, in a debris-littered patch of ground. (It would keep on chirping until technicians cracked it open and turned it off.)

Someone in the control room had shouted "Duck!" as hunks of metal began hitting the blockhouse, but Stehling didn't move. His back, which last gave him trouble when he was in an airplane with its engine on fire, was suddenly seized by a painful spasm. He stood still and watched the fire outside recede. When the smoke cleared, one section of the rocket, which had moments earlier been filled with dangerous fuel and poised to take off, hung off the side of the charred launch stand, looking like the skin shed by a snake. It was Dan Mazur, stubborn and blunt, who finally spoke:

"Okay, clean up. Let's get the next rocket ready."[14]

★

Upon hearing the Vanguard news, President Eisenhower was, in the words of his press secretary, "certainly" disappointed. Headline writers around the world apparently were not. In London, for example, they got the opportunity to cry "Flopnik" (*Daily Herald*), "Kaputnik" (*Daily Express*), and "Phutnik" (*Daily Mirror*) in the editions published immediately after the failure. Pictures of the fiery end of TV-3 appeared on TV newscasts across America. Hand-wringing broke out from coast to coast.

Amid renewed comparisons to Pearl Harbor, careful politicians refused to criticize directly the Vanguard program, its contractors, or actual managers. This would be like blaming American soldiers for a battlefield loss, and no self-conscious patriot would risk such a thing. Instead Johnson, Hubert Humphrey, Henry Jackson, and others put together a connect-the-dots kind of analysis that made embarrassment (rather than failure) the main issue and blamed either the Eisenhower administration or the press.

Johnson, speaking at a luncheon in Amarillo moments after the explosion, said, "I shrink a little inside of me whenever the United States announces a great event and it blows up in our faces." A moralistic Hubert Humphrey said he was "saddened and humiliated by the cheap and gaudy manner in which the Administration has gone about the business of trying a last minute rush to launch a satellite."

As Democrats complained that the Eisenhower administration had been wrong to hype the Vanguard launch, Republicans raced to the next dot on the picture—the press. Both Congressman Gerald Ford and the Pentagon's spokesman, Murray Snyder, used this deflection technique, saying that journalists should share responsibility for the nation's shame because they had built up expectations.[15]

The argument over *whom* to blame for Vanguard would continue for weeks in Washington while the community of technicians, engineers, scientists, and military officers who worked with the rocket searched for *what* caused the disaster. Herbert York got full-scale drawings of TV-3 and its inner workings and laid them on the floor of a hallway of the Old Executive Office Building in Washington. He and Vanguard director Hagen spent hours on their hands and knees going over the drawings in an attempt to understand what happened.

A similar exercise was conducted in Florida, where cleanup began as soon as all the little fires set by flaming debris were put out. On Saturday, December 7, the day after TV-3's spectacular death, the crew gathering the charred remains was interrupted by an Air Force Thor missile that roared off its pad at Launch Complex 17, Vanguard's next-door neighbor. It turned southeast, broke through the clouds, and disappeared from view on a flight that was shorter than expected but otherwise successful. On Monday the postmortem work was delayed a bit when, typical for the Cape, thousands were slowed for hours on their way to work. Rain and the usually heavy volume of cars were blamed for a cascade of collisions on Route A1A involving more than thirty vehicles.

Once they got to the missile range, the engineers and technicians performing the forensics found collected pieces of the rocket laid out in a hangar. Included were parts and fragments found by divers in the shallow waters close to the shore and surprisingly intact elements of the rocket, such as the third stage, the satellite, and high-pressure tank. As members of the team sifted through the debris looking for clues, they were puzzled by one strange-looking part that turned out to be a water pump from a car, deposited by a prankster to test their sense of humor.

Film, still photos, and recordings of the various instruments used to monitor the rocket were all analyzed. No one bit of evidence was conclusive. The readouts did not provide enough reliable data. The pictures, moving and still, showed events only from a distance. Future launches would be recorded by a camera placed much closer, but in this case the postmortem depended on wide-angle views that were subject to a number of interpretations.

Eventually the best theories on the cause of the disaster focused on the fuel tanks manufactured by the Glenn L. Martin Company, and how they worked with General Electric's first-stage motor. In talks that were sometimes heated, a consensus gradually emerged behind the idea that a fire started in a GE fuel injector and that improper design allowed the fire to spread into the fuel system. The engine-maker resisted this idea, and preferred a theory that blamed Martin workers for loosening fuel supply lines by using them like steps on a ladder as they worked on the rocket. Ultimately the official cause of the accident was declared "indeterminate." But the offending GE part was redesigned.[16]

The mood at the Cape turned gloomy as the postmortem proceeded. But Vanguard's failure didn't dim public interest in rocketry. Instead, newspapers and magazines published a stream of reports on space exploration. The most enthusiastic kids and adults took up model rocketry, buying ready-made rockets or putting together their own makeshift missile. A genuine rocket fad arose. It was popular in wide-open places like the Mojave Desert, but was also practiced in thickly settled Eastern cities and towns. By the U.S. Army's count, more than 1,400 missile clubs were created in the year after *Sputnik I* was launched. Although many, like the Arkansas City Technological Organization of Rocketry and Aero Ballistics, adopted names laden with grown-up-sounding officialese, their ranks were filled mainly by teenage boys who were amateurs in every sense of the word.

Some few, however, took the hobby beyond the model stage. In Queens, New York, a fourteen-year-old named Michael Tate, who attended the Bronx High School of Science, built a nine-foot-long rocket in his bedroom. Michael was advised by his neighbor Willy Ley, Wernher von Braun's friend who had fled the Nazis in 1935 and become a popular science writer in America. When it was finished,

Tate's rocket was deemed airworthy and capable of reaching an altitude of five miles. The trouble was, Michael had no launch site.

The same obstacle confronted a group of older boys—girls were deemed distracting—who created the American Rocket Research Society in nearby Brooklyn. They developed small, multistage rockets in an apartment- house basement littered with copies of the journals *Astronautics, Air Force,* and *Jet Propulsion.* Members of this group were so serious-minded that they hoped to sell rockets to the United States Weather Bureau. *The New York Times* praised them in an article called "It's Rockets, Not Rock 'n Roll, for Youths Who Dig for Facts."

The Brooklyn boys, who competed under the glorious name Flatbush Society for the Fabrication and Ignition of Pyrotechnic Projectiles, made home- based rocketry seem safe. It was relatively safe, as long as the rockets weren't fueled and fired. With firing though, accidents were common. For example, in the weeks after TV-3, sixteen-year-old Johnny Easley and seventeen-year-old Billy Hembree planned and built their own miniature Vanguard out of a two-foot copper tube, which they filled with a fuel made of zinc dust and sulfur. On ignition in a field outside San Angelo, Texas, the rocket seemed a dud, "just like the Flopnik," said Billy. When he and Johnny went to inspect their rocket, it suddenly exploded.

Billy and Johnny suffered minor injuries, and were released after brief hospital stays. They were lucky. In two incidents that occurred in the same season a teen rocketeer in Vancouver, British Columbia, died from exposure to fumes from fuel and a Texas science teacher was killed when his homemade rocket exploded.

Fortunately teachers could tap the rocket craze to inspire interest in science without risking their lives. In Merion, Pennsylvania, Arnold Gorneau's ninth-graders turned a class discussion about space travel into an elaborate study of the physical effects of the rapid acceleration men might experience as they rode a rocket into space. To start, a little bureaucracy was created with committees that researched everything from mouse physiology to nutrition. At the center of the effort was a research and development group assigned to build a machine that would create enough force so that a lab animal might experience stress equivalent to a rocket launch. Among the ideas considered and rejected were:

A bicycle wheel spun by hand with the aid of gears.
A cage tied to a rope swung by a student.
A tube attached to a vacuum that would suck an animal
    through it.
A rocket-powered roller skate.
A slingshot that would fire a tin can.

Ultimately they settled on a wooden disk spun by a one-quarter-horsepower electric motor to create a centrifuge effect. After a brief study of aerodynamics the class built a cage that looked like a piece of airplane wing, with a space for a small rodent. But as so often happens in big science, technical problems dogged the Merion centrifuge experiment. When the motor was first switched on and the disk was spun, the whole thing vibrated so violently that it seemed it would break apart. Braces were used to attach the table that the centrifuge occupied to the classroom walls. But then another test brought the scent of burning insulation and another emergency shutdown. Overheating was solved when the motor was rewired and compressed air was used as a coolant.

When all the problems seemed fixed, the Merion freshmen gathered to test their contraption. Two minutes before powering up the motor, they switched on the compressor and the cooling fans. Current was then sent to the motor, and it began to spin the disk, faster and faster. Then, as it hit top speed, the energy created by the spinning motor split the disk itself, sending pieces flying all over the classroom. Though not as dangerous as an exploding rocket, the hazard posed by flying pine begged an immediate solution. A new, oak disk was made and a wire cage was built around the entire contraption to make the next test safer.

The class worked for many weeks on the space travel experiment, eventually sending several subjects on simulated rocket trips and then studying their bodies and behavior. In a report on the lesson, their teacher noted that his students had learned much about experimental science and sub-specialties ranging from animal care to engineering, as well as the demands of teamwork. Unfortunately, Mr. Gorneau didn't leave behind a record of the experiment's results or word on the ultimate fate of the mice, except for the fact that there were more at the end of the project than there were at the beginning. During the course of the project two litters were born in the classroom, providing unanticipated lessons on reproduction. [17]

SIX

The Acid Test

Flashbulbs popped like Fourth of July fireworks as photographers captured General J. Bruce Medaris, in dress uniform complete with Bronze Star, Legion of Merit, and Distinguished Service medals, raising his right hand and swearing to Lyndon Johnson that he would speak, in the senator's words, "the truth and nothing but the truth." Behind him in the Senate hearing room waited Wernher von Braun, who would testify next and share top billing for the December installments of hearings that had become the season's biggest political show.

Like all good headliners, Medaris and von Braun had done their preparation. While the Vanguard group had struggled to launch TV-3, von Braun had labored with an assemblage of other rocket experts, military officers, and contractors to create a blueprint for a long-range American space exploration program. Made public one day before the Vanguard disaster, the plan called for a manned mission to the Moon and back within twenty years. Other exciting highlights included spacecraft suitable for carrying up to ten men and "research comets" sent to Venus and Mars. In one of the earliest recitations of an argument that would soon be a staple for space promoters, the group insisted that all the money poured into rocket science would yield invaluable new technologies and services for humanity, including heart surgery performed in weightless operating rooms.

Von Braun and his backers were respectfully silent on the day Vanguard imploded on its pad at the Cape. But before noon the next day the wizard of Redstone Arsenal was offering his views on spaceflight for New Yorkers listening to WRCA radio. (Coincidentally, perhaps, the station's owner, Radio Corporation of America, was a major space contractor.)

A few days later, the Army gave the press a photograph of technicians in Alabama fitting what was described as the final stage, the body of a satellite, on top of Missile 29. The picture, which ran on the front page of *The New York Times*, gave the public a first look at the Army's version of a satellite. Appropriately enough, since this was mainly a military project, instead of a friendly little silver ball with antennae like *Sputnik* and the Vanguard payload, the Army satellite—a cylinder pointed at one end—looked like the bullet that was going to finally hit the target for America.

When the photo was distributed, a spokesman coyly insisted the Army was trying to avoid the kind of publicity buildup that preceded Vanguard's debacle. Naturally, the picture had the opposite effect. It contributed to the notion that the Army was about to recover America's honor by putting something in orbit. It also helped build excitement around the upcoming testimony of Medaris and von Braun. Because they would speak at a rare Saturday session of the Senate, they were assured maximum attention from the press on an otherwise slow news day. In their years together, the general and the rocket man had enjoyed no better platform, no greater opportunity to push for their vision of America in space.

★

Senator Johnson opened the hearing with a grave promise to uncover the truth "in this time of peril" and praise for General Medaris as a "distinguished combat officer." The general was questioned by senators and committee lawyer Cyrus Vance. (Described that day by one senator as "the brilliant young assistant chief counsel," Vance would later become secretary of state.) For his part Medaris played the serious military man, soberly outlining the state of affairs. He reviewed the history of the Red-stone and Jupiter rockets and told how the Army team could have put a satellite in orbit in 1956, if only it had been permitted to do so. He complained about slow bureaucracies and restrictions placed on his ambition to build bigger rockets.

If the proceedings took on the quality of a scripted play—and they did—it was because committee lawyers had already interviewed the witnesses at length, and informed everyone involved on the topics and facts to be covered. When Medaris strayed too far or failed to hit his marks, Senator Johnson intervened, reminding him of what he had earlier said to the lawyers and coaxing him to "lay it on the line." While Johnson, another Senator who was coveting the White House, said the truth would be good for the country, it would also be good for the Democrats and his own future.

Eager to cooperate, Medaris caught the senator's directions and posed as a gallant defender of freedom frustrated by timid higher-ups who failed to recognize the value of long-term research and development programs. "We have got to develop engines and guidance systems and those things—even if we don't know what we are going to use them for today—because just as sure as they come through, by that time there will be plenty of use for them, and you have got to get down the road."

The more Medaris complained about the constraints put on him by the administration, the more Johnson liked it. Like a parent who encourages a child to "just tell me what's wrong" the senator stroked the general when he did well, praising him in a soothing Texas drawl, "You are getting down now to where you are telling us now what we want to hear, what we want to know, what we have to do, what we have to face up to," said the senator.

In the general's view, America needed to seize the highest ground possible for battles on Earth with satellites and, ultimately, space stations in the sky. "We need a ten- to twelve-year program that has, as its ultimate goal, the manned domination of space, and if we do not we are going to be in trouble."

Seated nearby, von Braun surely recognized that the general's plan for keeping the Army relevant in the space age matched his own lifelong dream of an integrated program of exploration that would include space stations, satellites, and journeys well beyond Earth's grasp. At last the man who had been arrested by the Nazi SS because he dabbled in space was hearing a forceful public argument, voiced by a prominent general, on behalf of all of his ambitions. When the talk got down to specifics, Medaris was blunt about what America's space projects needed most in the short term: more power.

Instead of incremental developments, said Medaris, the United States should commit immediately to matching the powerful Soviet R-7. A rocket with this kind of thrust—one million pounds—would bring parity when it came to nuclear weapons delivery and get a 1,000-pound satellite into space. "My personal opinion is that unless we develop an engine with a million-pound thrust by 1961, we will not be in space," said Medaris. "We will be out of the race."

In the meantime, of course, America would require an adequate force of shorter-range rockets to carry atomic warheads and Medaris left little doubt about who should wield them. It was perhaps the only instance in history when a witness before a Senate committee simultaneously claimed support for an idea from Soviet military planners and former Air Chief Marshal Sir Philip Joubert of the Royal Air Force. Medaris cited them both as he argued that missiles were artillery and should belong to Army ground forces.

★

Once the subcommittee and onlookers had enjoyed Medaris's warm-up, von Braun took to the microphone at the witness table. Dressed in a dark suit and white shirt with wide cuffs, he looked every bit the aristocrat. Later, one reporter couldn't resist describing the "youthful-looking German" as a "blond, broad-shouldered and square-jawed" performer who "made a hit" as he drew laughs and admiration from the politicians arrayed before him.

While his style impressed, as always, the substance of von Braun's testimony would satisfy those senators who sought an insider's critique of the Republican administration. Before he was finished, he would make it seem as if the administration's poor management had delayed not only satellite launches but deployment of nuclear-tipped rockets essential to Cold War deterrence.

As he fulfilled the committee's political purpose, von Braun also indulged his own need to promote space exploration. Noting the "mortal danger" America faced in the Soviets he proposed a new National Space Agency to be funded with $1.5 billion per year. This organization would draw on top scientists and engineers to develop satellites for weather observations, scientific research, and communications. He even suggested that a cluster of satellites could replace all conventional mail with electronic signals. But the agency's main tasks would involve manned flight. It would be charged with putting a human in orbit, and returning him safely by 1962. Next would come an attempt to control space in the way that a navy might control the seas, with an armed orbiting space station.

Like Medaris, von Braun talked at length about the problems of the military bureaucracy and described how it slowed his progress. With prompting from committee members and counsel, he described the dozens of overseers, agencies, and committees involved in approving rocket programs, assessing their progress, and doling out funds. At times the hearing room erupted in laughter as von Braun and chief special counsel Edwin Weisl teamed to poke fun at the bureaucracy. In a long give-and-take over the matter of visiting academic committees, advisory committees, Pentagon committees, and others, Weisl finally asked what happened to the plan for a million-pound-thrust rocket.

"In this area too, a committee was appointed."

Later in the day, von Braun would describe a future when Soviet satellites the size of *Sputnik II* might serve as spies watching America and as platforms for attacking U.S. forces and civilian sites with guided bombs. From the vantage point of space, these weapons could be steered to strike even moving targets, such as ships at sea. But such a fate could be avoided if American space programs were given the highest priority and provided ample amounts of money.

Von Braun dazzled the committee, whose members rushed to fawn over him. They worried aloud about whether he would become discouraged and flattered him with questions framed to show off his expertise. Mississippi Democrat John Stennis, ten years into a forty-two-year Senate career, asked, "What makes the satellite stay up? What keeps it on the move?"

The answer, worthy of a Disney presentation, boiled the principles of Earth orbit down to a stalemated contest between gravity and the satellite's velocity. Imagine a bullet fired from a position high above the rotating Earth, began von Braun. It travels forward and then starts to fall while the planet below turns away. If the bullet's speed is timed perfectly with the rotation of the Earth, the two will never meet.

The German-born expert's easy, authoritative response, coming at the end of his testimony, allowed the drama to end with a reassuring final act. In this vein, von Braun offered a vision of a future when America would be well defended and space science and exploration would yield great benefits for mankind. From space, experts might even make the heavens rain and alter the course of hurricanes.

In return for his fine performance, senators did what they could to let the world know that von Braun was, and always had been, a good guy pursuing noble goals under difficult circumstances. But even as they sought to paint von Braun as an idealist, the members of the subcommittee came close to evoking the darker side of his past. The hawkish Missouri Democrat Stuart Symington quizzed von Braun on the disparities between the 40,000-person workforce the Germans had dedicated to the V-2 and the 7,000 men and women engaged in Redstone Arsenal–related rocket projects. As von Braun started to explain who was included among the 40,000 deployed for the V-2 (among them were as many as 20,000 slaves), the senator quickly interrupted him.

Senator Symington: . . . Dr. von Braun, you testified last year that there were 40,000 people working on the V-2.

Dr. von Braun: Yes sir, including production and everything.

Senator Symington: How many have you working down in Huntsville?

Dr. von Braun: One hundred and twenty former V-2 men. But they were the cream of the crop. The 40,000 included—

Senator Symington: How many, all told, have you down there working on missiles?

Dr. von Braun: In my division, the development Operations Division of the Army Ballistic Missiles Agency, I have 3,500 people.

Aside from this one moment, von Braun and the Democrats on the committee were able to achieve perfect harmony. He was the expert man of science who helped them criticize Eisenhower and offered support for policies they preferred. They offered him appreciation and approval. This partnership extended to a proud exchange in which Chairman Johnson and von Braun agreed that the German missile experts who were brought to America at the end of World War II were better than the ones who went to the Soviet Union. According to von Braun, his old colleagues were contributing little to the USSR's achievements and were unhappy with their work and living conditions. In contrast, "I am very happy with the assignments I have been given in this country," added von Braun. "And I am very proud that I could make some contribution in this field so vital to our security."

"We are very happy to have you," answered the senator.[1]

★

While the Senate subcommittee entertained Medaris and von Braun, Dwight Eisenhower was in Paris for the first-ever meeting of the heads of states belonging to the North Atlantic Treaty Organization. Six doctors, including three neurologists, had declared him well enough for the trip, despite a standard for care that called for several more weeks of post-stroke rest. Looking robust and steady, Ike had stood at the doorway to the *Columbine III* and cheerfully waved his homburg before he departed on his transatlantic flight.

The president was annoyed by speculation over his health, especially suggestions that he retire for the good of the country, and he was determined to prove himself fit for duty. During most of the Paris summit he was able to keep up with the other leaders. Gathered around a circular table that was twenty-four feet in diameter and decorated with the NATO star symbol, the other heads of state beamed as Ike said, "Heroic efforts will be needed to steer the world toward true peace."

However, despite his show of vigor the president did sometimes struggle to speak clearly. In one twenty-minute address that was broadcast live over TV and radio he stumbled a dozen times over simple words such as "habits," "arrayed," and "decay." He also chose at the last minute to rest rather than attend one late-night dinner. Considering that his stroke was first explained to reporters as a "chill," the press corps raised concerns about the decision. Press secretary James Hagerty consulted again with his boss and returned to quote him, "Tell those gentlemen I'm a 9:30 or 10 o'clock boy tonight and I am going to bed at that time."[2]

The rest did Ike good and his health continued to improve. Focused mainly on Soviet advances in missiles and satellites, the NATO meetings led to an agreement to counter this threat by speeding the deployment of American intermediate range ballistic missiles in Europe. This move meant Eisenhower could immediately counter an attack by Soviet R-7s with IRBMs that could devastate dozens of cities in the USSR. In a bow to anxious allies, especially France, the Americans also agreed to seek talks with the Soviets on slowing or stopping the arms race. This result was a greater victory for Eisenhower than skeptics had expected when he had departed Washington.

As far as the president was concerned, intercontinental ballistic missiles—not satellites—were the grave issue and he seemed profoundly grateful to have NATO's help in maintaining the balance of Mutual Assured Destruction. At the closing session of the summit Eisenhower interrupted at the last moment to say, "Maybe a higher power would want us to bow our heads." The startled leaders immediately agreed and after thirty seconds of silent prayer, the gavel sounded and the summit was closed.

Eisenhower was driven to Orly Airport along streets crowded by French citizens who remembered his leadership during World War II. At a ceremonial terminal building called the Salon d'Honneur he made a brief speech, declaring that the summit had made "peace a little closer . . . the chances of war more remote." He then walked between two rows of uniformed Gardes Républicaines, their raised sabers reflecting bright floodlights, and ascended the stairs to his plane.[3]

Back on the ground in Washington, Europe's liberator and the world's peacemaker became, once again, Eisenhower the politician, referee, and national psychologist. The second *Sputnik*, the one now orbiting with the carcass of a dog inside it, had, in combination with Vanguard's impotence, done far more to spook the country than the first

Soviet satellite. The president had abandoned his initial posture of grandfatherly calm and clearly sought to answer the Soviets with an American spacecraft, but there was little he or his top men could do about reaching this goal, except referee the Army-Navy competition. In the meantime, space fever had swept through the political and defense establishments and it had grown hotter in the president's absence.

Within Eisenhower's own administration, the Air Force was angling to create a new space war department and a bomber that could fly 2,000 miles per hour. Pentagon chiefs were suggesting $1 billion in new spending on missiles. They also wanted to team with the Atomic Energy Commission to develop a nuclear-powered airplane—one that could span the globe without refueling—in three years.

For years experts had debated the notion of an airplane powered by an atomic reactor. Many considered it a technologically futile pursuit. The need for radiation shielding made reactors very heavy and in a crash the fuel that might be released would poison people and the environment. But after *Sputnik* flew, the Soviets, like pranksters shouting "Boo!," boasted that *they* would have a nuclear plane soon. This claim was followed by widespread rumors and finally a story in *Aviation Week* noting that the USSR already possessed such an aircraft. This wasn't true and never would be, but the report energized American military officers and contractors who wanted one too.

The Soviets understood that American defense contractors, scientists, and others who benefited from government weapons projects would naturally use the *Aviation Week* story to argue for more money. (Their scientists also lobbied their government for funding.) Certain members of Congress who wanted to seem tough on the enemy or hoped to bring federal dollars to their districts might also get into the act. The net result would be to divert attention and possibly resources from other, potentially productive purposes.

The chief scientific propagandist in the Soviet campaign of strategic disinformation was the same Anatoli Blagonravov who had beamed with pride at the National Academy of Sciences in Washington on the day after *Sputnik I* went into orbit. It was a demanding task. In December 1957 alone Blagonravov announced that Soviet scientists were making progress toward creating:

1. Power rays to transmit electricity great distances without cables;
2. Portable atomic power stations for the hinterlands;
3. A nuclear-powered train that could pull heavy loads at high speed.

These grand Soviet claims would be cited by Americans who sought money for their own research and ambitions. When Eisenhower got back from the NATO summit, he was greeted by a proposal for a ten-year, $10 billion program to develop space-related science and technology. This plan was offered by a group of more than two dozen scientists whose ultimate vision included human settlements in space. Given that the entire federal budget was less than $80 billion, the scientists recognized they were asking for a huge expenditure. As one justification for this sum they warned that the Soviets would invest mightily in their own space program to maintain their lead. Only a huge response would propel America past its foe.

Many other big ideas were thrown on the president's desk, and each one involved spending money. The National Education Association proposed a big increase in federal funding for schools—as much as $4.6 billion more per year—so that teachers could train the brainpower to compete with the Soviets and defend the homeland. The National Science Foundation called for tripling its annual grant budget for basic research projects on everything from biology to nuclear physics. Both of these groups would get support from Democrats and Republicans who would recognize investments in schools and labs as both good politics and good policy. But months would pass before any agreement on actual dollar amounts would be reached between the more cautious White House and congressional Democrats who pushed for a more aggressive approach.

Here again, Eisenhower would come in for some harsh criticism from Democrats. Backed by Lyndon Johnson and others, Arkansas senator J. William Fulbright likened Eisenhower to James Buchanan in the years before the Civil War, spreading "the contagion of his own confusion over the land." Fulbright wanted quick and dramatic action to stimulate new science. In addition to a White House plan for college scholarship aid, he suggested sending $500 million per year directly to public schools to spark a "true revival of learning."

Fulbright's attack, voiced on the floor of the Senate, was remarkable in two ways. First, he bucked long-standing and widespread public resistance to federal involvement in local education. Second, he pointed to a troublesome American tradition of anti-intellectualism "going back for many decades." In a nation where book learning was often met with suspicion, politicians more typically mocked academics and intellectuals. But in the post-*Sputnik* moment, as America's Cold War enemy seemed ever more threatening, old political restrictions seemed to lose their strength. It was possible to promote almost anything if it could be made to seem like a matter of national security.

Soviet propaganda and American fear caused some of the president's critics to revive old protests about his defense priorities. Many military officials had long opposed the president's reductions in defense spending and his peacetime preference for airpower over Army troops. Seeing an opening to revive the debate on policy, General James Gavin, the paratroop hero, publicly criticized the Pentagon hierarchy and then, weeks later, suddenly resigned.

A public drama that had that feel of something staged then ensued, with other generals expressing regret that this dashing celebrity general might resign and Secretary of the Army Wilbur M. Brucker offering Gavin another star if only he would stay. Gavin refused to be wooed and instead departed the Pentagon to write a book that would be part memoir and part manifesto for the American military in the space age.

Gavin's resignation signaled that America faced the most urgent kind of challenge, and many of the advocates, contractors, and experts who suggested expensive countermeasures spoke with similar passion. They were opposed by just a few quiet voices who challenged the assumption that an emergency was at hand. These cooler heads included men such as ARPA head Herbert York and Lawrence Kimpton, chancellor of the University of Chicago. Kimpton wasn't sure at all that American schools were behind those in the USSR and doubted that every child needed extra doses of science. York was almost certain that the quest for an atomic airplane, as well as other exotic weapons, was folly.[4]

But Kimpton, who spoke publicly, and York, who raised concerns within the government, were a decided minority. Almost everyone else in the public sphere used the space issue to stoke feelings of doubt and dread. In one case a high-powered married couple, writer-politician Clare Booth Luce and her husband, Henry Luce, publisher of *Time*, made this activity a family endeavor. First Clare used the annual Al Smith political dinner in New York to question America's moral and technological standing in the world. Then her husband made Nikita Khrushchev *Time*'s "Man of the Year" for 1957, an honor that received notice worldwide.

The *Time* cover portrait showed Khrushchev flashing a bad-teeth smile, balancing a tiny *Sputnik* ball on his fingertips and wearing the Kremlin on his head like a crown. The article depended heavily on the information that Soviet officials crafted for Western consumption, and as such had a peering-through-the-keyhole quality. *Time* had plenty of information about the chairman's public successes and the USSR's achievements. But American journalists could only guess at the problems that were hidden from view.

One big, but secret, crisis that had demanded much of Khrushchev's attention in the last part of the year was the growing environmental disaster following an explosion of nuclear wastes at Chelyabinsk. Occurring a week before *Sputnik I* was launched, the blast contaminated a wide area around the Soviet center for nuclear weapons fuel.

Of course, nothing about Chelyabinsk or other secret problems facing Khrushchev would be revealed by *Time*. Instead the magazine had to be content reminding readers that the Soviet Communist Party's chief was a tyrant and the principal enemy of the free world. *Time* also managed to paint him as a brilliant, earthy, impassioned, "overweight Will Rogers" who charmed Soviet farmers and workers while he reinvigorated communism worldwide. With the liberal use of foreign aid payments he made strategic inroads in the Middle East. With the *Sputnik*s he gained prestige in Africa. As one Ghanaian schoolmaster asked *Time*, "If the Russians are so oppressed, how could Russian talent be so creative?" Most significantly, the article made Khrushchev seem vital, engaged, and energetic at a time when America's president, in the magazine's telling, had been forced to risk his health in order to show a vigorous response.

Fortunately, for him, Eisenhower's gamble worked out well. In Paris he had shown himself to be capable and healthy enough to lead. At home he would try to demonstrate similar strength. First he resisted many of the efforts, including those within his administration, to bust his budget with big new programs in education, science, and defense. Having thought deeply about the last category—military spending— Eisenhower firmly believed that national security depended as much on moral strength and a solid economy as it did on weapons. "A point can be reached," he argued, "at which additional funds for arms, far from bolstering security, weakens it."

With these principles in mind, the president would begin work on plans to add to federal funds for scientific research and schools.

But where the immediate race for space was concerned he would have to hope that the established satellite programs would succeed soon. Here he found that one problem that had annoyed him for weeks was being addressed. Although both the Navy Vanguard group and the Army team were aiming for late-January launches, the Department of Defense had announced that the next satellite attempt would be made in secret. This policy might limit the celebration of a success, but Pentagon officials intended to spare the nation, and the president, the embarrassment of another failure witnessed by a flock of reporters.[5]

★

In December 1957, Cape Canaveral security officers began to restrict access to the rocket-watching areas near the launch site, closing off one road to the beach and forcing onlookers to walk an extra half-mile if they wanted to occupy the spot with the best view. They also started using helicopters to buzz the beach at moments when reporters and photographers seemed too close. The rotors kicked up a sandstorm, forcing camera operators to cover their lenses and, sometimes, to flee the beach entirely.

But even with the occasional helicopter sandstorm, and an official policy of secrecy, it was impossible for Cape officials to hide completely what was happening with various rockets. Three successful launches— a Thor, a Jupiter, and a long-range Atlas nicknamed Big Annie—were each witnessed and reported. Bigger than Jupiter, the Atlas was designed to become America's first intercontinental ballistic missile and Annie's flight was one of the most impressive witnessed at the Cape to date. According to one account, the wife of an Atlas worker named Jerry watched Annie from the beach on December 17 and prayed, "Oh God, please make it go! Help Jerry make it go right!" That night, after Annie went right, booze flowed freely at the Starlite Motel lounge. Over scotch and soda the manager for the missile's main contractor, a self-described "slave-driving bastard" named G. G. McNabb, declared Big Annie "majestic."

A few days later, at an airstrip on the Redstone Arsenal in Huntsville, Alabama, a front-loading Air Force cargo plane called a Globemaster gobbled up Missile 29 like a shark swallowing a fish and then flew it down to Patrick Air Force Base. Crews there transferred the missile, which was to be named after Jupiter's wife, Juno, onto a truck trailer and then shrouded it with canvas. It was moved to the launch center before dawn, traveling along Route A1A through the center of Cocoa Beach, past Bernard's Surf restaurant, souvenir shops, and all the motels. Anyone who asked about the rocket would be told that it was a regular old Redstone, nothing special. Certain people associated with the missile would also travel in secrecy, taking military transport to the air base and avoiding publicity.

Although special precautions might limit the public's awareness, they could not hide the Army rocket team's prelaunch activities at the Cape from its Vanguard competitors. For one thing, the Vanguard people who worked a few hundred yards away to repair the firing structure at their own launch complex had a perfect view of Hangar D when Juno, queen of the gods, arrived in her canvas shawl and was rolled inside. For another, they were fully aware of the schedule for launchings at all the Cape complexes. They knew that the Army was gunning for a January 29 attempt, followed by another one on March 5. (General Medaris had pushed hard for this second date, just in case his first try met the same fate as TV-3.) If the Navy was going to beat the Army in this game, they would have to move very quickly. The sight of their intramural competitors "added to the general scurrying," recalled Kurt Stehling.[6]

While they rebuilt their burned-out launch facility, the Vanguard team was hampered by a severe cold snap that began the worst season in memory for local citrus growers, who would lose 30 million boxes of fruit. At the windswept Cape, nighttime temperatures often went below freezing and during the day laborers needed extra clothing. Managers bucked themselves up with reassuring chats about the difficulties of rocket science—it *is* as hard as brain surgery—and the halting progress nearly all rockets made from design to reliable flight. The Thor that flew so impressively before Christmas, for example, was the descendant of many failures, including three launch pad explosions and a midair breakup. One perfectly executed Thor launch was actually aborted on a ground controller's command because someone misloaded special paper in a printer that plotted the rocket's performance. With the paper upside down, positive readings looked like a disaster in the making and the man at the controls hit the self-destruct button.

Inside Hangar S, where the Vanguard workers assembled pieces of their next rocket, they found electrical problems that required many repairs on connections and calls for new parts. Checks of the fuel system and rocket engine also turned up more items that had to be repaired on the spot. Finally, on a cold day when indecisive clouds produced a steady mist but no real rain, the first stage of the rocket was cleared for a test firing on the rebuilt pad. For ten hours a crew fought electrical shorts caused by the drizzle until everything looked right. When they finally lit the engine, it fired for all of three seconds before a monitor saw what looked like a bad reading on a faulty gauge and threw a switch to turn everything off.

And so it went, through the first weeks of January. A second test was also delayed by equipment failures. Then, when everything seemed ready, the countdown for the so-called static-firing, when the engine would be fired with the missile fixed in place, was halted because someone noticed that a hatch door on the side of the rocket wasn't properly closed. A gantry crane that had been rolled away had to be put back in place (a twenty-minute operation) so that a technician could reach the door and tighten it. The engine was then test-fired and an apparently faulty reading on a gauge in the blockhouse forced a premature shutdown.

For Vanguard engineer Stehling, the rocket began to seem like a beautiful but temperamental thoroughbred horse. Potentially powerful but delicate, the slender thing presented one challenge after another to trainers who loved her, but lived in a near constant state of frustration. The test firings had not provided conclusive proof of the rocket's abilities, but the crew had to proceed as if Vanguard would be ready to run when called upon. Still in line for launch ahead of Juno, Stehling and his colleagues had the chance to make history.

With round-the-clock effort, they were ready to try again by the third week in January. At 9:20 P.M. on Wednesday, January 22, another cold wet night, the countdown clock was started with hopes for liftoff by eight the next morning. A crowd of engineers, managers, and bureaucrats from the Naval Research Laboratory and the Martin Company had come south to witness the firing. Their appearance was one of many clues that led locals to conclude that a satellite shot was imminent.

For the first time, however, reporters were alerted ahead of time and brought by bus onto the test range to observe. Air Force officials had built an observation deck on the flat roof of a building and equipped it with a small covered section of bleachers and a dozen or so plywood booths with phones that would be turned on after launch. Under an agreement with General Donald Yates, the reporters would maintain a news blackout until the launch was completed.

Young Wickie Stivers was on the first bus that took reporters from a gathering spot at Patrick Air Force Base to the new press site. At first the prospect of watching a launch up close instead of straining to see it from the beach was enough to keep her and the others focused and excited. But as hours passed and the dark night grew colder, the opportunity began to feel more like hardship. The greasy burgers and bad coffee offered by the women who ran the food trucks were little comfort, especially since the only toilets available were chemical closets with no running water.

"You were definitely stuck out there, and there was no option of leaving," recalled Wickie, years later. "Most of us had parts of our stories, the background, already written. But we had to be there for the most important part of the story—the lead—which depended on what happened with the missile. You had to be there to grab the phone and send out a bulletin."

For Wickie and other local reporters, these bulletins could mean a very big payday because they had contracts with stations around the country that paid as much as $50 apiece for timely news flashes. Wickie's clients included stations as far away as Philadelphia and Washington, D.C. This was how space reporting acquired a Southern accent that it would keep for years to come, as many of these early local stringers became employees of major networks.

Through the night the reporters endured cold, wind, and then rain as they watched for any sign that ignition was imminent. At the launch center, where the rocket was illuminated by powerful lights, nervous men wandered between the lunch trucks and the blockhouse, where the air was thick with the smell of burned tobacco and perspiration. Inside, a telephone line was kept open to the office of Vanguard director Hagen. Except for odd moments when the often overloaded local phone system caused a disconnection, the link would be maintained for the entire countdown.

The countdown clock was first stopped at 10 P.M., just forty minutes into the prelaunch dance. A two-hour delay was required to fix a small part on the rocket. When this chore was finished, another problem—this one fuel-related—caused further delay. To pass the time, the visiting dignitaries stared over the shoulders of the men at their control stations and eavesdropped when groups gathered to discuss fixes for these problems. The most sensitive might have heard in the tone of the voices a little extra anxiety as the group shared an unsettled feeling about their prospects for success. Only much later would some admit that they had premonitions of failure.

When they weren't coping with mechanical problems, the Vanguard team worried about the wind, the rain, and later even hail that fell just before dawn. Still the countdown continued. Spirits rose in the morning as the weather cleared and forecasters predicted about twelve hours of dry, calm conditions. Then, as the clock approached the four-minutes-to-firing mark, the satellite riding atop the rocket suddenly stopped talking to radios set up to hear its beacon. The countdown was halted. The gantry was wheeled back into place beside the missile and a technician scrambled up to discover that rainwater had shorted out the radio. The moisture was mopped up, the radio beacon was restored, and aluminum foil was wrapped around the nose cone to keep its cargo dry.

But even as this latest problem was fixed and it seemed a launch was just moments away, the experienced members of the crew began to fear their long night's effort would yield nothing. Their faces puffy with fatigue and dusted with whiskers, they peered through the thick, blast-proof window to see that ice was building up on the Vanguard's flanks. Caused by the super-cold liquid oxygen that helped fuel the rocket, the ice was so thick that it would alter the rocket's flight if it were launched. The launch was canceled.

The liquid oxygen was dumped out of the rocket and allowed to evaporate into the air. Managers decided to try starting again at 1 P.M., hoping to fire at 4 P.M. Since so many bugs had been shaken out of the rocket during the first try, they figured three hours would be enough time. And besides, the window of clear weather that had opened in the morning was expected to close soon.

The pressure these men felt as they struggled to reach space was substantial. Since no fully built Vanguard had ever flown, the crew couldn't call on experience for reassurance. They were, instead, like rookies suddenly called up to a major league team to play in the World Series. And like rookies, they were equal parts determined, nervous, and excited. All these feelings intensified as the clock raced against clouds that began to lower on the Cape.

In this period, a series of brief delays pushed the launch time back. Most painful was one that occurred when a tracking radar suddenly lost the ability to see the missile. In the precious time it took to realize that the blindness was caused by fog and was not a serious problem, the weather got the upper hand. At 7 P.M., with nine minutes to go in the countdown, the range safety officer intervened to suspend the action. The overcast was just too thick to allow a launch. Exasperated men groaned and grimaced. Some left their chairs to stretch. Others moved to the floor, where they slumped against walls. Once again the launch was canceled. The space reporters, who had been awake and attempting to stay alert for twenty-four hours, were bused back to Patrick Air Force Base.

The missileers and journalists who drove home quickly got six or seven hours of sleep before returning for another try at a Vanguard launch. On January 24, a Friday, they got within twenty-two seconds of firing when a switch that was supposed to disconnect the "umbilical cords" that supplied the rocket with gases and electrical power failed. While the team inside the blockhouse waited, two men drove a cherry picker truck up to the rocket and one rode the boom up to fix the switch. For a moment the reporters posted nearby thought that the technician was being loaded into the nose cone for a ride.

Once the cord release was fixed, the countdown resumed with the clock reset to thirty minutes. Again the seconds ticked to below thirty and then twenty. In the blockhouse the tension built until it felt like the men inside could throw the satellite into space with power of their collective anxiety. Then, at fourteen seconds, a valve that was supposed to close a vent that released excess liquid oxygen froze. Several attempts at a repair failed; with the valve still venting the oxygen mist, a pall swept the blockhouse. This was the last mechanical step before ignition, and it could not be taken. Another launch was canceled.

By this time, liquid oxygen had been loaded into the rocket and then discharged three times. Engineer Stehling, part of the propulsion group, now began to worry about another combustion source, the acid-based fuel that helped propel the second stage of the rocket. This corrosive stuff had been on board for days and leaks were a definite possibility. In a conference with representatives of the engine's manufacturer, GE, Stehling was told that Vanguard's tanks could contain the liquid and his concerns were set aside. The fuming nitric acid, as the fuel was called, would remain on board and another countdown would begin at 1 P.M. on Sunday, January 26.

The foreboding that Stehling had been feeling for days was confirmed minutes after the clock was switched on that Sunday afternoon. At about 1:15 a worker who had been tinkering up high on the rocket suddenly screamed and then slid down an emergency pole. Others rushed him to an emergency shower and blasted him with water. Behind them, halfway up the rocket, a brown cloud began to emanate from the second stage. Acid was either leaking or was on fire.

In an instant, fire crews led by Chief Norris Gray shooed every nearby technician away and turned their hoses on the rocket and the pad. The blockhouse team quickly learned that the injured man had been sprayed with acid when he opened a small hatch to make an inspection. Further investigation uncovered a small leak that had allowed enough acid to collect to cause this danger. Besides injuring the worker, the liquid had destroyed an important weld inside the missile.

This was no fog bank, hailstorm, or frozen switch. It was a significant problem. They could either attempt a complicated, time-consuming repair or cannibalize a waiting missile for a replacement engine. For ten hours crews inspected the rocket and launch leaders debated their next step. Even Vanguard chief Hagen, in Washington, was consulted by phone. All the while they felt the presence of their competitors, von Braun and team, hovering nearby like a "ghostly consortium." The Army's target day for launch was January 29, just three days away.

During the hours of debate, additional acid leaks were discovered and a repair job was ruled out. Day turned to evening. In the dark winter sky, at last cleared of clouds, the dog coffin that was *Sputnik II* passed overhead at 6:22 P.M., reflecting flashes of light from the sun, which had just set in the west. Alone in space since *Sputnik I* had reentered the atmosphere and disintegrated in early January, it was bright enough to be seen clearly with the naked eye.

No one at the Cape wanted to utter the words that had to be said as the rocket's fate became obvious. Finally, in Washington, Hagen took the lead. The second-stage engine would have to be replaced, he concluded. This operation would require several days. According to the schedule set by Air Force test range managers Vanguard would have to yield to the Army and its queen, Juno.[7]

SEVEN

Being Nonchalant
and Lighting Up a Marijuana

After so many months of resentment over Vanguard, *Sputniks*, and other setbacks and insults in the struggle for space primacy, General Bruce Medaris's opportunity had arrived. At his headquarters in Huntsville, where two huge and muscular uniformed guards always stood outside his office door, spirits rose. The general and his team were so confident that they began planning a victory party for their satellite launch more than a month before the date when their satellite vehicle—Juno—was supposed to be fired.

February 1, 1958, would mark the Army Ballistic Missile Agency's second birthday, and Medaris was certain that his crew would have "a real achievement to brag about." He meant that James Van Allen's instruments would be safely in orbit by then, courtesy of a Redstone-made rocket. Thus, the anniversary would also be a satellite celebration, and perhaps a national coming-out for Huntsville, the rocket city that America tended to forget.

Because all the big blastoffs and explosions happened at the more glamorous Cape, Huntsville and the ABMA never attracted throngs of reporters. And unlike the sunny palm-lined coast, northern Alabama was hardly a national tourist destination. Sure, the region had its attractions, like the Tennessee River and the state park atop the highest local peak, Monte Sano. But it was best known as a stopover for travelers on their way to Florida from points north and west. Just a decade earlier Huntsville was still so rural and old-fashioned that a guidebook published by local boosters made no mention of aerospace or sophisticated manufacturing but did feature a photo of a farmer working a team of horses above the caption, "Plowing makes a man feel good."[1]

Outsiders could be forgiven if they considered Huntsville a sleepy nineteenth-century kind of place, but for locals any sense that their hometown of 15,000 was behind the times had disappeared when the Army rocket group and its Germans arrived in 1950. They helped to reopen a pair of World War II arsenals and establish a sprawling complex for designing and making rockets. The 40,000-acre military base and the contractors who came to serve it created thousands of jobs that paid far more than the handful of manufacturing companies that were the big employers in the past. Huntsville was a rocket city, whether the outside world knew it or not.

By the end of 1957 the annual payroll for the rocketworks approached $100 million a year and a typical employee earned about twice the national average income of $3,700 per year. The cash flowing out of the arsenal in the form of paychecks and contracts transformed the city. New housing developments—ranches with big windows were popular—sprang up as the population doubled and then doubled again to 60,000. Downtown Huntsville and its courthouse square, guarded by a statue of a Confederate soldier, bustled. On Monte Sano the Germans began building a colony of beautiful homes. On the west side of town, modern commercial strips grew along the new Memorial Parkway, which collected the heavy traffic headed to Redstone every morning and delivered it back to the new subdivisions in the evening.

Sputnik had all but guaranteed that the Redstone Arsenal and Huntsville would thrive for many years to come. Indeed, in a spasm of Cold War muscle flexing, Congress had started the year by approving the president's request for $1.2 billion to accelerate all missile programs and then adding $40 million specifically for Army rockets. In this same period Washington lifted the 200-mile restriction of the range of Army rocket weapons, and authorized work on a completely new missile to be called the Pershing. Together, these actions would stimulate more hiring, more contracts with vendors, more growth for the local economy.

But since the arsenal was a military facility working only on classified projects, this new money was slow to change Huntsville's image. Though proud of their prosperity, local leaders understood that Rocket City received far less positive publicity—the kind that attracts tourists and new businesses—than the Cape region. Even worse, many Americans looked askance at places like Huntsville, where blacks lived mainly in what one writer termed "squalid slums," simply because they were part of the South. It was easy for outsiders to see the entire region as the home of conflict, racism, and fear.

Segregation and black poverty were pronounced in Huntsville, but the city was also comparatively peaceful. Granted, blacks were not yet getting the good new jobs, but given the influx of engineers, scientists, and businesspeople with more open views, even their future was brighter. Huntsville was in fact a leader in technology and science, and a good showing by Juno would bring local boosters the chance to show this side to the press and the world. It would be good for the city's spirits, good for its reputation, and ultimately help diversify the local economy making it less dependent on the arsenal.[2]

★

Anyone who read the local paper, *The Huntsville Times*, knew that a Jupiter C had been prevented from putting an object in orbit more a year ago. The editors left no doubt that the local rocket boys were the best bet for America's space ambitions. Two days after Vanguard's disaster at the Cape they published a big, front-page photo of a Jupiter C—one just like Juno—above the headline, "Jupiter C Fever Spreads to All Corners of Nation." The fever was so high in Huntsville that the leaders of the local chamber of commerce handed a Distinguished Citizen Award to Wernher von Braun six weeks *before* he would take his big shot.[3]

Kudos and credit were a big concern in the inner sanctums of ABMA too. In the weeks leading up to the Juno launch, General Medaris focused much of his attention on the public relations plan for the event. Still eager to separate military projects and science done in space, the Eisenhower administration seemed determined to hide the Army's role and play up the contributions of scientists such as Van Allen and civilians at the Jet Propulsion Laboratory. Medaris feared being passed over by the national press, and devoted much effort to making sure that wouldn't happen.

Negotiations with Washington eventually produced a formal publicity strategy so complex that Medaris would compare it with the D-Day invasion plan. It laid out rules for public information officers, print reporters, broadcasters, and photographers, and set conditions for the release of background material and key bits of data. As background information was prepared for the press, Medaris noticed that just about everything that might show his group in a good light was stricken for national security reasons, while every civilian with even the slightest involvement would be welcomed center stage.

According to this public relations plan, in the event of a successful launch the National Academy of Sciences would play a big role as a source of information on the satellite's orbit and payload. Every effort would be made to describe the satellite as an instrument of peace and progress. At the same time, officials of the Jet Propulsion Laboratory in California would discuss tracking reports and the performance of the rocket's upper stages, which the lab had designed as a system of rockets on top of rockets to create the necessary thrust for orbit.

The setup for the upper stages relied on smaller rockets that had been used, with consistent success, for several years. Eleven of these rockets were arranged in a ring to make up the second stage, which would fire once the main Jupiter rocket ran out of fuel. Three more small rockets were fastened together to comprise the third stage, while the fourth and final stage was a single small rocket that would go into orbit as part of the satellite. Ignited in proper sequence, all this firepower would create the necessary speed—roughly 18,500 miles per hour—required for the final stage to break free of Earth's gravity and enter orbit.

With JPL bragging about its three stages and the scientists explaining the satellite, what could the Army group talk about? They would be confined to answering questions about the main rocket. But because it was a piece of classified military hardware, their answers would be brief.

Although Medaris and his publicity chief, Gordon Harris, lost almost every skirmish in the PR planning process, they still tried to position their group for attention and praise. They decided that as the launch approached, Wernher von Braun would camp in Washington and join William Pickering of JPL and James Van Allen at the Pentagon War Room, where teletype messages, projected on a big screen, would provide constant information.

Von Braun would miss the flames and thunder at the launch pad, but bask in the flash of the news cameras at a follow-up press conference. It would be up to General Medaris to represent the Army at the Cape, and this was one of the main duties he had in mind as he departed for Florida on Monday, January 27. By this time the Vanguard group had determined that its one available rocket had been so damaged by leaking acid it was no longer airworthy. Vanguard wouldn't get another chance for a week. In that time, the Cape would belong to the Army.[4]

★

At seventy feet, Juno was just half as tall as the Statue of Liberty without its base, and for more than ten days she had cast a shadow over the efforts of the Vanguard missileers, standing watch over their struggle and failure. It was hard for them to escape feeling intimidated, especially when they considered that Juno's old technology was also the most reliable around. Years later, Vanguard engineer Kurt Stehling would still describe the rocket as if he was talking about a human rival who "stood nearby, almost insolent, backed by a good history of testing and an eager staff of experienced missile engineers."

The Juno team's experience showed, even when it came to bureaucratic gamesmanship on the Cape. In talks with missile range commander Donald Yates, Medaris managed to win a guarantee that his team would have extended and exclusive support for a launch once Vanguard scrubbed. The Army would have no competition for services until Juno flew, or the launch was canceled.

In response to President Eisenhower's frustration over the public humiliation that came with TV-3's demise, Yates also set up the kind of press arrangements that would guarantee worldwide publicity for the satellite launch under conditions that were so well controlled they would please even the most paranoid image-maker. Yates was a friendly, small-town fellow from Maine with a bright, toothy smile. His charm helped him to keep the press in line, but when it failed him, he only had to remind reporters that he controlled all the things that they wanted: accurate information about missions, access to the launch pad area, and knowledge of launch schedule.

According to the Yates plan, which *Time* magazine dubbed "peace-time censorship," reporters and photographers would be alerted hours ahead of time and bused to the test area where the view, a little more than a mile from the launch pads, was far better than it was at the beach. There they would be tended by press officer Ken Grine, and would receive extensive background briefings and printed materials filled with accurate information.

Grine was perfect for this job. A space enthusiast and career military man, he would protect the image of the space pioneers. But he also liked journalists and wanted to help them. "They were pushy, but they had to be," he recalled. "Getting them out to the Cape solved a lot of their problems, but it created others."

The main problem was isolation. In exchange for their special access, the journalists at the launch site had to agree to keep silent until after Juno was fired or retired. In this age before wireless communication, Grine controlled the phones in the wooden booths at their viewing area, which was built atop a structure called the Optics Building. The phones wouldn't be turned on and available for reports until after Juno was fired and everyone involved was ready for the news to go out.

With the news media corraled and an experienced team following an established routine to launch a reliable rocket, everything that could be controlled was. But there was nothing the Army or the Air Force could do about the jet stream. Soon after he arrived at the test range, General Medaris learned that the winds at altitudes from 25,000 to 40,000 feet were gusting up to 200 miles per hour. At issue was not the strength of the winds, but rather, the difference between the jet stream force and the winds in the adjacent layers of the atmosphere. Too sharp a division and a rocket could be damaged or even destroyed as it passed from one zone to the next.

Weather balloons were sent aloft and the data they collected was then transmitted from Cape Canaveral to the ABMA computing center in Huntsville. The response was not encouraging. According to protocol, Juno should not be launched until the upper atmosphere winds diminished substantially or, better yet, the jet stream shifted away from the area. But there was some good news. The Cape was on the very southern edge of the stream and the winds would have to move only forty or fifty miles northward for a launch to be possible.

For three days the launch team could only sit and wait as meteorologists continuously monitored conditions aloft. The Army group could depend on the reports and analyses they received. After all, the Cape's forecasters included commanding General Yates himself, who had provided critical weather predictions for Eisenhower's invasion of Nazi-occupied France in 1944. Their expertise was unquestioned, but conditions on the ground were so pleasant—sunny, calm, warm—that it was hard to believe the upper atmosphere was too turbulent. At sunny Patrick Air Force Base, reporters who had been told a launch was imminent pestered their minder, who could only say that buses would roll when forecasters gave the okay.

Because they had refrained from fueling the first stage of the rocket, the Juno group didn't have to worry about corrosion and damage to the machinery. But those involved in the launch were getting worn down by anticipation. The hours of overtime duty logged by range workers were expensive, and then there were all those in Huntsville, at JPL in Pasadena, and in Washington, D.C., who were anxiously awaiting the launch.

Medaris thought about all the important people with a keen interest in the launch and about the celebration planned for ABMA's anniversary on Saturday. He hoped for a break in the weather. Finally on Thursday the 30th meteorologists discovered that the jet stream was beginning to move north. The winds in the upper atmosphere decreased slightly. Medaris consulted with his launch team leader, a German V-2 refugee named Kurt Debus. Even those who liked him thought that Debus, whose face had been scarred in a fencing match, could be imperious and condescending. But his high standards meant that he could be decisive when necessary. He advised Medaris to start the countdown and aim for ignition at 10:30 P.M.

Working from a long list, the crew inside the blockhouse began running checks on the rocket's electrical and mechanical systems. Engineer Robert Moser, who oversaw the countdown, would recall that the air in the bunker was filled with tobacco smoke (his brand was Chesterfield) and anxiety as he and Debus hovered over a team that was so experienced it could perform with very few direct orders.

Not yet thirty years old, Moser had spent much of his childhood in Pennsylvania but graduated high school in nearby Daytona Beach. Gifted in science and math (he appreciated the precision of disciplines like physics), Moser had studied electronic engineering on a scholarship at Vanderbilt. Drafted into the Army during the Korean War, he was eventually assigned to Redstone Arsenal, where Medaris recognized his talent and quickly turned him back into a civilian so he could concentrate on missile work free of a soldier's other duties. Tall and skinny with an exceedingly calm demeanor, Moser wore glasses and except for a strip in the center of his forehead, his hair had receded halfway back on his skull.

True to his nature, Moser had developed an artful and precise system-by-system checklist that would be used to make sure the Juno rocket was in good condition prior to ignition. Beginning with the order for the ground crew to remove tape from the rocket's valves, he would spend hours calling off the chores to be done.

Early in the countdown, most of the action took place outside on the pad as workers prepped the rocket. Then, as the count progressed, the five key controllers in the blockhouse had more and more to do. On Moser's signals they would throw the proper switches, check gauges, and look for warning lights. At some of these stations, TV screens showed color pictures of the rocket outside. In a calm moment, Moser could leave the small group and walk to one of the thick, green-tinted, blast-proof windows that faced the pad and get an even better view. But these breaks were few.

"There was a sequence we were supposed to follow and that was more important than the actual time being counted down," recalled

Moser decades later. "There really were no significant problems with this vehicle. And if something needed to be fixed, someone could walk right out to the rocket and work on it."

In fact, while it sat on the pad the rocket was controlled by simple electrical relay switches or mechanical valves housed in a space beneath the pad that was accessible down a few steps. When a problem arose, a technician could scamper out, duck down into the little cellar, and put his hands right on the problem. Moser got a kick out of walking to the pad when liquid oxygen was being pumped into the rocket. Excess would continuously cascade down the side and turn to vapor, making the men on the ground feel like they were walking in a cloud.

The tension that came with the countdown, the satisfaction of troubleshooting problems, and the anticipation of what might happen when a rocket was fired, all thrilled Robert Moser. Married just a month earlier (a quick honeymoon was spent in Havana), Moser missed his young wife during long shifts at the Cape. But this feeling faded when he was busy and the countdown was moving forward. Like many others, he could stay on duty for twenty hours or more, depending on adrenaline, caffeine, and nicotine to maintain his focus. And no matter the outcome—success, exploding failure, delay—he felt like he was part of a something extraordinary, a key player on a great team of explorers.

With a little more than four hours to go, as Moser worked down his checklist, buses that had been loaded with press people at Patrick Air Force Base rolled through the security gates and then on to the viewing area. So far the reporters had obeyed the terms of the agreement they had made with General Yates. While the public had been informed that work was underway for an Army assault on space, specifics about the time and date of the launch were held secret.

The reporters understood that Medaris and Debus had bet on a change in the winds aloft. At dusk weather balloons were released and they carried instruments into the upper atmosphere. The probes found the winds had decreased, but not substantially. The numbers were sent by Teletype to the awkwardly named Aeroballistics and Guidance and Control Center at Huntsville, where they were fed into a computer-based flight simulator. While they waited, the launch team got ready to pump liquid oxygen into the rocket to complete the fueling. Finally, a little more than an hour before launch time, the analysts in Alabama sent word via Teletype.

"Highly marginal—we do not recommend that you try it."

History would have to wait another day. Medaris and a few colleagues left the Cape for a drink and some sleep.[5]

★

The information embargo imposed by General Yates worked. When the launch was called off, there was no repeat of the TV-3 drama, with big headlines stoking public interest. But still, anyone who had paid attention had a general idea that the Army was pushing for a launch by February 1. This much had been reported. And it would have been easy for anyone hanging around the bars and coffee shops near the Cape at the end of January to detect the influx of contractors and government bigwigs who accompanied first the Vanguard attempt and then the revving up of the Juno effort. The flow of traffic was changed because workers lingered past their shifts. Motel vacancies dried up. Fathers were suddenly absent from family outings.

If they had observers in the area, and many missile men assumed they did, the Soviets were also alert to the signs of an impending launch. And you didn't have to be paranoid to suspect the Kremlin was tweaking the American side when information on plans for a big new *Sputnik* was suddenly and mysteriously leaked to Western sources. According to so-called experts outside the Iron Curtain, a monstrously large rocket was ready to go and its keepers were poised to launch the biggest satellite ever, perhaps in a matter of hours.[6]

The news that the Soviets might be in a countdown greeted Medaris as he awoke on the morning of January 31. In fact, no satellite launch was imminent. But Medaris couldn't know this when he made his first check on the weather and got good news. The stubborn jet stream had finally budged, shifting far enough to the north that the winds in the upper atmosphere would recede to less than 120 miles per hour by nightfall. The routine of checking out the rocket, tracking facilities, and communication lines would be repeated on this day, with a formal countdown beginning at 1:30 P.M. Once again members of the press would take a late afternoon trip to their isolated viewing area, and Kenneth Grine would make sure they behaved.

Fatigue and excitement competed inside everyone involved in the preparations. Medaris fought against exhaustion by taking a few brief naps during the day in his private cottage at the Trade Winds Hotel in the small community of Indialantic, about twenty-five miles south of the test range. Unlike other rocket vagabonds, who stayed in the motels closest to the Cape, Medaris traveled the extra distance to the luxurious resort, which had been built at the height of the Roaring Twenties and offered a golf course, pool, and fine restaurant. He was a celebrity there, and received especially attentive care supervised by the owner, Tom Doherty, who had bought the neglected Trade Winds after the Great Depression and restored its grandeur.

Still confident in Juno, Medaris believed he would need his rest because a successful "shoot," as he termed it, would be followed by a press conference that might last until 3 A.M. (Ever the ordnance man, Medaris's use of the word "shoot" over, say, "launch" emphasized that missiles were a type of ground-based weapon used by the Army, rather than aircraft that might serve the hated Air Force.) Medaris had also scheduled an early morning flight to Huntsville for Saturday. City leaders were already planning a hero's welcome at the airport.

After one last report on the weather was delivered to his room at the Trade Winds, Medaris put on his uniform, heavy with medals, and departed the hotel for the Cape. When he arrived there at 6:30 P.M. he learned the countdown was running just five minutes behind. In the blockhouse Medaris found Moser wearing a headset and a microphone that was held by a wire around his neck and rested on his chest. He trailed a telephone cord as he paced behind the stations where the five key controllers sat.

Clipboard in hand, Moser had worked down the familiar checklist with hardly a halt for almost six hours. He was happy to see Medaris arrive. Others may have found him abrasive, but the general had always treated Moser well. The younger man had shared the general's frustrations about Washington's attitude toward the Army rocket team. He also shared the general's confidence that Juno would carry America into orbit and break the Soviet monopoly in space.

Moser stood or paced around the blockhouse as he issued pre-launch commands and fielded responses from the controllers. (Later he estimated that he "walked off a quarter inch of shoe leather" with his pacing.) At 4 P.M. he paused to eat a grilled cheese sandwich, which would have to fuel him through the night. With an hour and a half to go before ignition, Moser ordered that the self-destruct devices on the rocket be armed, so a safety officer could blow it up if necessary.

"Arm destruct block," he said into his microphone.

"Roger, destruct block armed," came the answer in his headset.

A hose that delivered hydrogen peroxide to the rocket was automatically disconnected. A dribble of liquid prompted Moser to call for a brief hold. A technician ran out to the pad and looked under Juno's motor and found that the fluid was just excess and did not come from a leak. (While this was being checked, Medaris stood with Debus, nervously smoking a cigarette.) About thirty minutes later, as chatter stopped and the blockhouse grew still, Moser exhaled loudly, said, "Phew!" and grinned to break the tension.

A phone rang. Officials at Patrick Air Force Base, which was twenty miles away, wanted to know who would relay word of the launch as it occurred. Medaris, who was watching from a spot out of the way of his team, said that he would handle this duty.

With twelve minutes to go before launch, the lights in the blockhouse were dimmed so that Moser could better see the rocket through the thick glass of the window. A minute later Moser called to the pad safety officer to make sure the area outside was clear of people, equipment, and other hazards like food trucks. With the pad area safe, he then turned to a controller named Terry Greenfield, who threw the switch to start an electric motor that would spin the upper stages of the rocket. This spinning would create centrifugal force to keep the payload on track as it flew. Out on the pad, the electric motor whined, signaling the news reporters at their distant post that Juno was coming to life.

The red-painted service tower was rolled away from the rocket. Juno stood alone on the illuminated pad, looking like a slivery, opalescent fish in the moonlight. A loud siren began to send a deep-toned warning across the Cape. A Teletype was sent to the Pentagon:

WHEN THE COUNTDOWN REACHES ZERO, THE
BIRD WILL NOT BEGIN TO RISE IMMEDIATELY, SO
DON'T BE WORRIED IF WE DON'T TELL YOU IT'S ON
THE WAY.

Safe inside the blockhouse, Debus and his deputy moved to the window, where they joined blockhouse chief Moser. Within a few minutes everyone in the room was standing, hoping to catch a glimpse of the rocket on the pad in between turning winches and checking meters, gauges, and lights. One of the men in the room looked to Medaris and said the countdown was going more smoothly than previous ones.

With five minutes to go, Debus lit another smoke.

With less than two minutes to ignition Moser heard a troubling report in his headset. A small problem had occurred with a vane on the rocket. He looked to Debus for guidance. The German with the scar on his face waved him off. "Go ahead!" he said.

Moser, calm as ever, counted off the final ten seconds out loud. At "one" he said, "Firing command."

The order began a rapid-fire series of events. First pumps began to build the pressure of fuel flowing to the rocket motor. Then special igniters lit the fuel. Initially the fire was invisible, as it built up heat. When the right temperature was reached, valves that controlled the flow of fuel opened and the big rocket engine started to rumble. In the control room, flashing lights indicated that everything was working properly.

This firing process took, by one measure, almost sixteen seconds. Then an orange flame began to glow at the rocket's base and a roar rumbled across the Cape. As a cloud of smoke billowed out in every direction, the rocket climbed on a bright flame. At first the rocket moved so slowly it seemed it would sit back down on the pad. But once it was about a hundred feet in the air, Juno picked up speed and climbed faster. Accelerating steadily, it raced toward the stars, growing smaller in the eyes of those who watched from below.

"Go baby!" shouted one of the control room crew. A few others applauded and shouted "Go! Go! Go!" A message was sent to the War Room at the Pentagon: "IT'S SOARING BEAUTIFULLY."

Outside, on the roof of the building where their plywood phone booths waited, the assembled members of the Fourth Estate looked to the sky and gaped. A few miles away, on Bird Watch Hill and down on the shores of Cocoa Beach, thousands of people craned their necks to watch the rocket ride its flame into the dark night. Smaller crowds, including high schoolers whose parents worked at the Cape, stood watching on the beaches near Titusville, where they could see the launch pads and feel the roar of Juno in their stomachs.

At Jaycee Beach, where perhaps two dozen high schoolers had camped to watch the launch, JoAnn Hardin's mind and heart were filled with a range of ideas and emotions. Like her friends she experienced the thrill of the event and intense patriotic pride as the rocket rose and then disappeared behind the bank of clouds. But she also knew she was witnessing a scientific achievement of historical importance. Her idols included Galileo and the great European explorers, and she knew that this rocket and satellite stuff marked the beginning of a new age of adventure, discovery, and technological advance. She felt lucky to be alive.

In the moment, of course, the cheerleader in JoAnn Hardin pulled for Juno's success in the way that loyal fans root for their team to win a game. Just about everyone who watched the launch shared this intense desire. But no one followed the rocket's course with more intensity than General Medaris and Debus, who immediately turned to the instruments that tracked Juno's flight. The first stage exhausted its fuel supply in less than three minutes and it automatically separated itself from the rest of the rocket. After about four minutes of flight, when the first stage had dropped away, a signal was sent to ignite the cluster of eleven second-stage rockets. These fired properly, and then the third and fourth stages burned automatically.

With the pad area safe, the blockhouse door was unlocked and the men inside, who had viewed events from the worst possible perspective, could emerge for some air and a brief glance at the sky, where their creation had already disappeared.

Desperate for confirmation that their most important rocket was performing as it should, Medaris and a few others got into waiting cars and sped to a communications and tracking center set up in a nearby Quonset hut. There a team of analysts led by JPL's Al Hibbs was already racing through calculations based on the apparent speed and angle of satellite's flight. Hibbs was a slightly famous math whiz with a quirky personality. As a college student he and a friend had studied the mechanics of certain roulette wheels and figured out how to beat the casinos in Vegas. The stunt got them into *Life* magazine and earned them enough money to buy a sailboat and finance an eighteen-month cruise around the world.

Hibbs would recall that an anxious Medaris was "having kittens" as the math team analyzed Juno's flight with scanty data. Normally Hibbs would have had help from a tracking post on the Caribbean island of Antigua, but when the crew there flipped the switch to transmit their information to the Cape, nothing happened. (Later it would be discovered the switch was corroded.) Without these numbers, Hibbs had only the rocket's speed, and the time it passed over the horizon to use in his calculations. With so little to go on, he told Medaris he was "95 percent" sure that the satellite would stay aloft.

"Don't give me any of this probability crap, Hibbs," said Medaris. "Is it up?"

"It is up," answered Hibbs, with assurance.

It was fortunate that Medaris consulted Hibbs and not the range safety office about the status of the satellite. The safety experts had done separate calculations that indicated Juno may have failed to develop adequate thrust. According to their math, the satellite would circle the globe on a shallow path and come crashing down somewhere near Tampa, just a hundred miles or so short of a complete orbit.

Since he was oblivious to the conflict between the calculation being performed by Hibbs and the safety office, Medaris called Patrick Air Force Base to announce Juno's success. At about this time Hibbs was flooded with powerful feelings. He went outside and stared into the sky. As he struggled with his emotions—relief, accomplishment, pride, even a sense of disappointment that the long satellite struggle was over—he found himself crying alone in the darkness.

Like the Soviets with their *Sputniks*, the Army team would have to wait for their satellite to complete one full circle around the globe before they could know with absolute certainty they had accomplished their mission. If the satellite, Deal, did indeed come sailing around to

American airspace, it would undergo an immediate name change. The new name, which was needed for public relations purposes, had been debated at the highest level. With a nod to sophisticates of the day, General Medaris had suggested "Highball." The secretary of the Army favored "Topkick." But the president had chosen, in the end, the simple name *Explorer.*

The wait, especially as the satellite crossed the vast Pacific, where no listening posts had been established, was difficult for everyone involved. Nerves drove the worried and the weary to chain-smoke and drink too much coffee. At the Cape this ritual was made a little more pleasant by an official who had supplied novelty mugs decorated with the image of a Jupiter rocket. At JPL in Pasadena, humor broke the tension. When Medaris sent a message instructing them to take it easy for a while, they answered that they were "BEING NONCHALANT AND LIGHTING UP A MARIJUANA."[7]

★

At the Pentagon, where von Braun, Van Allen, and Pickering waited with others in the War Room, no marijuana was available to ease the wait for tracking stations to report hearing the satellite's radio beacon. The key listening posts were in California and were connected with JPL. If they heard Explorer's radio signal at the proper time—roughly two hours after launch—that would mean that the satellite was indeed positioned in space, at a proper altitude, and the mission could be declared a success.

In an exchange that would later become widely known, Secretary of the Army Wilbur Brucker sent a playful telex from the War Room to Medaris: "I'M OUT OF COFFEE AND WE'RE RUNNING OUT OF CIGARETTES. WHAT DO WE DO NOW?"

Medaris replied: "SEND OUT FOR MORE AND SWEAT IT OUT WITH THE REST OF US."

Brucker, former governor of Michigan, told the men in the War Room: "This is like waiting for precinct returns to come in."

By 11:30 P.M. radio reporters at the Cape—Wickie Stivers included—had used the telephones at the press site to inform the world that Juno had lifted off and America's latest satellite launch was, at the very least, going better than the Vanguard try. Following Grine's instructions, the press then boarded buses for the theater at Patrick Air Force Base, where they would get more information. Minutes later several Army cars raced from the launch area out the gates of the test range and down A1A to the base. In those cars were Medaris, Debus, Hibbs, and a few others from the Quonset hut.

At the base theater, more than a hundred reporters were in their seats when Medaris and the others walked to the front of the auditorium. Medaris and Debus leaned back against the stage and casually shared what they knew, while they awaited confirmation of Explorer's orbit. Hibbs posted himself nearby, while a colleague, Charles Lundquist, went to a phone booth to contact the tracking network. He would stay on the line until he had news. Posted between Lundquist and Medaris, Hibbs heard the general tell reporters, "Don't worry, we'll hear from it."

Thanks to the reporters in the wooden phone booths, word of the successful firing had already spread from continent to continent. Amateur radio operators began reporting that they heard *Explorer*'s signals. Van Allen and the others in Washington didn't give these claims much credence, since the satellite's broadcast frequency had not been announced and it was not one commonly tracked by so-called hams.

Knowing that *Explorer* required more than 100 minutes to make a full circle around the Earth didn't make the time pass any more quickly. Confined at the Pentagon, Wernher von Braun had missed the fiery launch of the rocket that represented his first real chance at his lifelong dream of conquering space. Ever the politician, he had dutifully accepted his role as the Army's rocket program representative in Washington, giving up the chance to be with his team at the launch. Now, as the clock ticked away, he had to rely on the JPL group in Pasadena for words of reassurance. They didn't come as expected.

At the 100-minute mark, William Pickering was on the phone to officials at JPL, who had nothing to report. The same was true at 110 minutes into the flight. By all estimates, the California post should have captured the satellite's radio signals. Von Braun's spirits were shaken as he thought "obviously we were mistaken. The *Explorer* had never really gone into orbit." Nearby, Van Allen would notice that no one was speaking and "an air of dazed disappointment settled over the room."[8]

Almost a thousand miles to the south, Van Allen's assistant, George Ludwig, experienced a similar sense of foreboding. Having arrived at the Cape too late to get a spot inside the blockhouse, the young satellite-maker had found a haven inside Hangar S, where a temporary tracking station had been established. Since all the listening posts that were supposed to follow *Explorer* were in direct communication with each other, Hangar S would get the news as soon as it was available. But by 12:40 A.M., Ludwig and the others still had not heard word from California.

Finally, at about 12:45, Ludwig heard someone shout that a West Coast outpost had confirmed receiving *Explorer*'s signals. Almost immediately, other stations began reporting that they also heard the high-pitched tones broadcast on the two frequencies used by the satellite's transmitters. At 108 megahertz, radio operators heard alternating notes that sounded like the chirps of a mechanical bird. Nearby on the radio dial, at 108.03, a steadier, single note also hummed to declare *Explorer*'s presence in the sky. (The *Express* of London would describe the sound as a "continuous *wheee* that signaled, Cheer up America. We're in the heavens, all's right with the world.")

At the same moment that Ludwig heard confirmation of *Explorer*'s success, Charles Lundquist came out of his phone booth at the Patrick Air Force Base theater and flashed an "OK" sign at Hibbs, who walked quickly to Medaris and whispered the good news in his ear. As the general turned to the assembled press and announced this big success, the hall erupted in cheers.

Simultaneously, in quiet rooms scattered across the country, serious men who had braced themselves for disappointment heard the news and burst into celebration. At the Pentagon War Room, where William Pickering finally announced that JPL had heard the signals, generals and scientists and engineers pounded each other on the back and grinned in celebration. They realized that Juno had developed extra thrust, so that *Explorer*'s orbit was much higher than expected. This orbit, which ranged from a low of 227 miles to a high of 1,575, accounted for its late arrival over California, but also assured it would resist the pull of gravity and remain aloft for years.

At the White House, press secretary James Hagerty called the president's getaway cabin at the Augusta National Golf Club, where Ike was beginning a brief vacation. He had left Washington only hours before, after giving a rather lackluster pep talk to the Republican National Committee, which was preparing for the fall election.

Of all the men who were waiting for the satellite news, the politically beleaguered president may have had the best reason to get excited. Throughout the evening, as he played bridge with his wife, Mamie, and friends, Ike had received reports on Juno relayed by the White House. At 10:40 P.M. a phone line had been opened from Washington to Augusta and it stayed open until the satellite orbit was confirmed. But with *Explorer* as with *Sputnik*, his response was measured. At first he said, "That's wonderful," but then he quickly added, "Let's not make too great a hullabaloo over this."[9]

Before 1 A.M. the White House issued an official statement to the press noting the satellite success, stressing its scientific value and ignoring the Army's role in the event. President Eisenhower also left the Army out of his formal congratulatory message, which was sent to the National Science Foundation.

Eisenhower's slight so wounded Bruce Medaris that he would call it the "most bitter disappointment of my life." But it wouldn't deprive the Army of public attention. At 1:30 A.M., Van Allen, von Braun, and Pickering took the stage at an auditorium at the National Academy of Sciences with two photo-friendly props: a full-size copy of the *Explorer* satellite, which weighed less than twenty pounds, and a model of the Juno rocket, labeled in large letters, "US Army."

The room at the Academy of Sciences was "jammed to the rafters, in the middle of the night, with TV and movies and everything," Pickering would later recall. In a press conference that stole the spotlight from a similar event at the Patrick Air Force Base theater, von Braun, Van Allen, and Pickering answered for more than two hours. A photo of them at the press conference, holding the model of *Explorer* over their heads, was published in newspapers around the world.

In the reports filed by journalists who attended the news conference, the Army would be credited with this "moment of triumph." (*The New York Times* paired this phrase with yet another reference to the "ill-fated" Vanguard.) And considering the reaction in Congress, where *Explorer* was hailed as a near miracle, General Medaris and every other member of the Army team should have felt duly appreciated. Lyndon Johnson declared the satellite "a tribute to loyalty, determination and persistence against great odds."

But mixed with all of the praise, leaders of both political parties also offered warnings about complacency and the need for a sustained space exploration program. America's prestige was somewhat improved, but the nation had not quite matched the USSR. As the Soviets pointed out, in a bit of macho posturing, their satellites were much bigger. These factors were likely behind Ike's restraint and modest coverage of the event in the press, compared with the public response to the two *Sputniks*. America had entered the space race, but was still far behind the leader.

Among those who failed to follow the president's low-key example were the citizens of Huntsville, who couldn't restrain themselves. Even before the White House made it official, the exciting news about the satellite spread from inside ABMA headquarters to households across the city. At the urging of the local radio stations and police officers making bullhorn announcements from their patrol cars, people flocked to the city's courthouse square. There hundreds joined police vehicles and fire trucks with sirens wailing in a parade around the Confederate monument and the Madison County courthouse. Eventually a crowd of thousands would gather in the square, where people cheered and wept, and Redstone workers were suddenly celebrities. Among the signs people waved were some that read, "Move Over Sputnik, Space Is Ours" and "Our Missiles Never Miss."

As the local paper would soon point out, in the wake of *Explorer* the community was due for more jobs, more federal money, and more prestige. Add these expectations to the sense of patriotic pride that Juno ignited and the people of Huntsville felt overjoyed. However, even in this high moment when Bruce Medaris, Wernher von Braun, and 4,000 Redstone workers had achieved one of the most impressive feats of the scientific age, the sense of grievance some of the revelers felt toward former Defense Secretary Charles Wilson, whom they blamed for holding back the Redstone group in the past, resurfaced. There, in the dark Alabama night, while horns honked and others shouted gleeful congratulations, they burned Wilson in effigy. It was a strange and ugly display, but Wilson didn't win much empathy the following day when he dismissed *Explorer* as a "good technical trick." He had said roughly the same thing about *Sputnik*.

Hours after the cars stopped circling the courthouse square and the happy citizens of Huntsville had gone to sleep, the plane carrying Bruce Medaris, conquering hero, landed at a nearby airstrip. There was no welcoming crowd of official greeting, as had been planned. Civic leaders had decided at the last moment that everyone, Medaris included, would be too tired to enjoy it.

Later in the day a combined birthday party and satellite celebration for ABMA drew between 15,000 and 20,000 people to the arsenal. General Medaris stood to review a military parade, and an Army-made film on the Juno-*Explorer* launch called X *Minus 80 Days* was premiered. But the highlight of the day was the test-firing of a Jupiter rocket engine that sent smoke floating high over Huntsville and thunderous noise across the countryside.[10]

EIGHT

A New Era of Exploration

At first Abigail Van Allen, busy with a houseful of children, thought the telegram that arrived on the Saturday night after *Explorer* went into orbit was another note of congratulation for her suddenly world-famous husband. But then she opened it and was surprised to find an invitation to dinner at the White House on Tuesday. Immediately she realized she didn't have the right clothes for a state dinner and, worse, the stores would be closed until she had to depart early Monday.

Iowa City being the kind of place it was in 1958, word of the invitation and Abigail's delightful problem quickly spread. Neighbors and friends soon arrived with dresses—more than a dozen—along with jewelry, gloves, and furs. On Sunday Mrs. Van Allen wobbled upon stacked volumes of the *Encyclopaedia Britannica* while a committee of friends helped her judge one dress after another. Eventually they chose a full-length gown of blue-gray taffeta with a modest scoop neckline and three-quarter-length sleeves. Her husband would rent his uniform of white tie and tails in Washington.

At the dinner, which was a buffet of little sandwiches, cakes, and champagne, the Van Allens joined a room where many were first-time guests of the White House and felt both excited and privileged. Many were likely new, also, to the entertainment for the evening, a performance by the singing comedienne Anna Russell. Gifted with an operatic voice, a mischievous mind, and perfect timing, Russell's parodies included *How to Write Your Own Gilbert and Sullivan Opera* and a thirty-minute version of Wagner's entire Ring Cycle.

The state dinner became a treasured memory for the Van Allens, who knew that as in grade school, party invitations—especially those sent by the White House—signaled a certain status and popularity. Though the event was formally a celebration of both the military and scientific leaders behind *Explorer*, the president had favored scientists with twice as many invitations. Conspicuously absent was General Medaris, while big names in physics, chemistry, and biology attended along with the famous Wernher von Braun.[1]

Von Braun was a diplomatic sort of scientist. On the night when Juno placed *Explorer I* in orbit, he told the press at the Academy of Sciences that the competing Vanguard rocket was in fact a superior launcher and had been ill-fated only because it was so new and sophisticated. In contrast, he added, the Jupiter C launcher was a reliable but "obsolete" machine based on old technology. The implication was obvious. Once the people wrangling it learned how to get Vanguard into the air with regularity, it would set the standard for a new generation of missiles.

Vanguard had been built with some advanced materials and designs. But von Braun's generous words were as much a policy statement as a technical assessment. Vanguard was widely viewed as an all-American (as opposed to German-American) product, and despite its Navy connections, it had been promoted by the president as a civilian project. Although Ike and the Congress had, officially, just begun to address the future of space policy, insiders understood that the matter of who would control the future exploration of space was almost settled in favor of the civilians.

Just weeks after *Sputnik I* went into orbit, Ike had created the President's Science Advisory Committee and named his new science adviser, James Killian, to lead it. The PSAC, as it was called, had met almost continuously ever since to develop two major policy proposals: consolidating control of military missile work in one office at the Pentagon, and creating a civilian agency for the scientific exploration of space. These ideas reflected the president's growing esteem for the scientific community and long-standing concern about the ambition and influence of military leaders and contractors.

The reorganization of defense missile programs had been included in the president's State of the Union address, which he had delivered in January. In the same speech Eisenhower had stressed the value of science in the service of peace, pledging that American discoveries would be shared with the world. In this spirit, he would soon begin working with Congress, and in particular Lyndon Johnson, on the long-range plan for space exploration. While certain generals, like Bruce Medaris, would fight hard to keep space a military province, a betting scientist would put money on the prospect that aside from weapons, all the real rocket work was going to be run by civilians.[2]

Having spent his adult life seeking sponsors for his dreams, von Braun was yet again positioned well, especially when compared with Medaris. At forty-five he was ten years younger and still in his vigorous prime. More smoothly comfortable with civilian bureaucracies, he had not made the kind of enemies that Medaris had in the scientific and political communities, and thanks to Walt Disney and others, he enjoyed wide public support despite his Nazi past.

Von Braun's popularity was all the more remarkable considering that at the very moment of Juno's success, four of his V-2 comrades were being prosecuted in Germany for the slaughter of 208 men, women, and children in the waning days of World War II. The massacre had taken place during the evacuation of V-2 units in the face of Allied advances. Von Braun and others who ultimately came to the United States were part of that retreat. But accounts of the trial in the American press did not connect the crimes to the German missile men of Huntsville. Instead, Wernher von Braun was widely hailed as an American hero and even pictured alone on the cover of *Time* magazine with a rocket seeming to blast off from his square shoulders. He was the man dubbed most responsible for *Explorer*'s success by *The New York Times*. And in the months to come, when the press, Washington's power elite, and business leaders sought an authority on space, they went to him.

In interviews and speeches von Braun insisted that even with *Explorer*, America had not matched the Soviets. "It will require years of concentrated effort to come abreast, and even longer to pull ahead," he said in one typical talk. Like all the men (and the few women) engaged in big science, von Braun had obvious interest in keeping the nation on edge about the competition with the Soviets. But what he said was true. As if to emphasize the point, the supposedly superior Vanguard suffered perhaps its most painful defeat in the dark hours of the morning after the scientists and military men had left the White House with their heads abuzz with champagne.[3]

★

Originally scheduled for ten o'clock on the night of Tuesday, February 4, 1958, the fifth attempt to launch TV-3 Back-up, and its small satellite, was marked by a noticeably relaxed mood. Juno had put *Explorer* in space and there was no longer an intramural race to be won. Vanguard's engineers and technicians had accepted the loss and gone back to thinking of their rocket as an experimental launch vehicle with a scientific purpose. John Neilon, who worked for the project at a radar station south of the Cape, would be glad the Vanguard team could focus on its original mission, launching a satellite for the International Geophysical Year with an advanced, more efficient new rocket.

This attitude allowed the Vanguard team to be calm and deliberate that Tuesday as they dealt with a fuel leak that required more than four hours to fix. (A tank was drained, fitted with a new O ring gasket, and then refilled.) The repair, conducted as temperatures dropped to the freezing mark, pushed the projected blastoff to 2 A.M. The only other significant delay came at one hour before launch when a safety officer noticed two fuel trucks—potential bombs in the event of an explosion—had been left near the pad. The count stopped while a truck was sent to haul them away.

In the quiet of the blockhouse, every light and gauge and meter said the rocket was ready to go. Missile-tracking radars were set and operating perfectly. Twenty minutes before launch, floodlights were switched on and the long, white rocket glowed brightly on its pad.

At a little after 2:30 A.M. the count reached zero, the firing order was issued, and the main engine on the slender rocket roared to life. In a moment it began to rumble up, off the pad. After just ten seconds and less than a thousand feet of flight it tilted toward the southeast, as expected, and began to accelerate. In the cold dark sky the light from the rocket's 6,000-degree flame competed with the glow of a full moon. Half a minute passed, and the men in the blockhouse could not contain their glee, as Stehling reported:

> *Everyone cheered and hurrahed. [Daniel] Mazur shook his head unbelievingly and ran to the window to look out. We tried to follow it as long as we could. [Robert] Gray led a charge into the telemetry room. Even the usually imperturbable [launch team member William] Escher was holding his hands over his head and dancing around. We had done it.*

One minute into the flight and approaching an altitude of 20,000 feet, Vanguard's first-stage motor had consumed more than half its fuel and by all signs was performing perfectly. Then suddenly a telemetry specialist named Mason Comer shouted above the din of celebration.

"Something's wrong here!"

In front of Comer the screens and ink plotters that tracked the rocket's flight had stopped showing its ascent. Men in the blockhouse turned to the windows and caught the sickening sight of what looked like a meteor blazing across the sky with a long tail. Vanguard had broken up and was falling to Earth. For a moment the main fireball seemed headed straight back to the Cape. On the flat roof where the assembled press watched, someone shouted, "Look out, it's coming our way!"

The pieces of TV-3 Back-up fell, like twinkling bits of fireworks, into the ocean about three miles offshore. They splashed into the water not far from John Neilon's radar station, where he was the sole Vanguard technician working with an Air Force radar crew. Neilon had patiently waited to see if all three stages of the rocket fired. "To borrow an old saying from mathematics, getting it off the ground was necessary, but not sufficient," he would note many years later.

Perhaps because of this, the failure struck Neilon as hard as it did the men at the blockhouse, who, according to Kurt Stehling, found the sight of the explosion "heartbreaking." Neilon would actually consider the flight, which was, after all a "test," a kind of success. "It did better than the first one," he would say. An investigation of the failure soon found its cause: a broken wire or electrical connection.[4]

Vanguard's failure knocked the good news about *Explorer* off America's front pages but it was not alone in its shame for long. The very next day the Air Force lost a massive Atlas, the nation's only true intercontinental ballistic missile, after it flew for less than a minute and then simply exploded in a puff of smoke. And one month after the Atlas, von Braun, Medaris, and company fell short in their attempt to put a second *Explorer* loaded with more of James Van Allen's equipment into orbit. They had hoped to tie the Soviets in the number of satellites launched.

The Army failure had a strange, lost-in-space quality. After a near perfect countdown, the launch motors were fired up at about 1:30 P.M. Crowds on nearby beaches heard the rocket growl as it burned huge amounts of fuel to lumber off the pad and watched as it flew on a copper-colored flame through the thick clouds that covered the region.

After the glow in the clouds faded and then disappeared, tracking devices showed the flight went well, with every stage firing properly. Then, when it came time for more distant listening stations to report *Explorer II*'s radio signal, the airwaves were quiet. With no clear sign that the rocket failed, but no reassurance from the satellite's radio beacons, the Army group fell silent and kept the press waiting. After several hours, an official confessed that the mission was in trouble but left open the possibility that the problem was with the radio equipment or atmospheric conditions and the satellite could yet be in orbit.

For nearly a full day the Army refused to confirm it had failed. But gradually it became clear that the final motor had failed to fire and the entire fourth stage—engine, satellite, casing, and Van Allen's instruments—had fallen from an altitude of 200 miles into the Atlantic somewhere near the island of Trinidad. The White House issued a statement saying that the president was "disappointed."[5]

★

Ever since the December flop, when he demanded the press covering space be tamed, the president had not expressed similar disappointment over Vanguard. When fuel leaks caused the cancellation of the launch in January, he said nothing. And after the Army won the race to orbit, he made sure Vanguard's chief, the physicist John Hagen, was on the guest list for the White House celebration. Ike was still favoring the civilians, even though they had so much trouble flying right.

Scientists who spent time with Eisenhower, including James Killian and Herbert York, would sense in him a growing appreciation for long-term investments in basic research and the development of new technologies. In the post-*Sputnik* era, when advances in physics, chemistry, and other disciplines were ever more vital to national security and prosperity, he relied heavily on the men he called "my scientists," and he expressed great confidence in them. He also seemed to accept that error and failure were inherent to their work. At times it was the president who urged patience on them, insisting that the Soviet threat was not so urgent that America was in immediate peril.[6]

This trust was extended to the civilian-dominated Vanguard project, even as rumors of its demise spread to the press. In this time, as he felt almost overwhelmed by demands for explanations—and apologies—from Capitol Hill, John Hagen received no pressure from the White House. The same was true for the men and women at the Cape, who labored in the hangar to prepare another test launch, with satellite attached. As John Neilon would eventually say, "We weren't the highest priority at the Cape, but we always got what we needed."

The Vanguard group was based in Hangar S, where rockets were assembled in a large open area flanked by workshops on the first floor and, above those, offices accessed by stairs leading to balconies. The federal government occupied the rooms on one side and the contractor Glenn L. Martin Company used those across the way. During the tense period when they had carried the nation's Cold War satellite hopes, Vanguard managers were especially nosy about the contractor's activities. Conflicts grew more frequent. Once *Explorer* had gone into orbit and Vanguard was no longer under pressure to match Sputnik, morale improved a bit, but there were still jibes thrown back and forth about the "piece of junk" missile and the inefficiencies of the federal bureaucracy. Everyone was tired of pouring time, energy, money, and emotion into a thing that refused to work.

With the new rocket called TV-4 being put together like a giant, three-dimensional jigsaw puzzle, a March launch was planned. Fog and rain delayed a planned firing during the first week of the month, frustrating the crew, the press, and various officers and bureaucrats who came in the hope of witnessing history. Once the weather cleared and they could start a countdown, the launch was ruined when a hose that delivered helium to the rocket refused to fall away when a disconnect switch was thrown. A call for a cherry picker truck to help solve the problem went unanswered. Its union driver, having received no order to stay, had gone home. The gantry was laboriously rolled back into position so the hose could be fixed. The delay allowed more foul weather, including heavy rain, to roll in, and the launch had to be canceled. The next attempt was foiled by so many technical problems—leaks, malfunctions, even missing parts—that the usual sense of hope and anticipation in the blockhouse was replaced by resignation and gloom. New rockets were generally expected to fail often as they were being developed, but the experience was nevertheless grinding and dispiriting.

It being the age of highballs, Stehling joined a group of Vanguard men who soothed themselves with alcohol at the Sea Air Motel. They ruminated about revolution but had trouble figuring out who would be the target of their rebellion. As the night wore on and the drinks added up, the subject turned to murder and its various means, and the denizens of the Cape who might deserve such a fate. But eventually their outrage ebbed and the drunken rocketeers determined that one more failure would at least result in a sternly worded memo to someone, which some of them would sign.

Fittingly, since drink, prayer, and luck were so much a part of early space exploration, St. Patrick's Day was selected as the next launch date. Feverish work was required to replace, tighten, adjust, and align all the pieces of rocket that were out of whack. (In a display of prayerful superstition, one worker secretly attached a tiny medal of St. Christopher, patron saint to travelers, to the second stage. This act would cause a minor church-state controversy later.) In the midst of all the stressful work, the tension was broken when a wife's surprise visit to her engineer husband turned up a Vanguard secretary in his motel room. It was a thoroughly amusing distraction for everyone who wasn't directly involved, as was an incident at a local bar involving a technician and the shattering of a row of bottles filled with expensive liquor.

The countdown began on March 16, with the hope that the missile would be launched after sunrise the next morning. Through the night, as the big checklist was followed, something extraordinary occurred—there were no significant problems. At dawn a clear blue sky greeted the crew. At about 7 A.M. a delay was imposed so that *Explorer I*, which was passing overhead, could disappear over the horizon. No one thought the two spacecraft would collide, but the technicians wanted to avoid confusion over radio signals and radar tracking.

When the count resumed and reached the four-minute mark, the local radars were locked onto the rocket, and listening posts in the Caribbean reported they were ready to pick up satellite signals. As per standard procedure, the ducts that brought fresh air into the control room were closed to prevent smoke from entering in the event of a fire or explosion. Immediately the temperature in the blockhouse rose.

About two minutes later, the fuel line that wouldn't detach during the last attempted launch got stuck again, but only briefly. It fell away after a few seconds and the two technicians responsible for this part did a little dance in the blockhouse.

The last half-minute of every launch seems to pass slowly and this was the case on March 17. Those who weren't required to monitor the rocket's systems at the consoles crowded around the window. Across the test range, at distant hangars, office buildings, trailers, and shops, people stood outside with their eyes trained on the pad. Kurt Stehling, who had chosen to post himself in a communications trailer instead of the blockhouse, stood outside, pressing binoculars hard against his face and staring at the missile.

When the command was made to fire the rocket, the smallest flame was ignited at its base. For a moment Stehling feared it would go out, like the flame on a camp stove smothered by a sudden breeze. But it didn't go out. Instead the flame spread and grew and the first-stage motor began sucking huge volumes of fuel and burned brighter and brighter until, after just a few seconds, it began to push the tall, thin body of the rocket up toward the sky.

All of the power—28,000 pounds of thrust—was devoted to trans-porting a metal ball that weighed less than four pounds and contained just two little radio transmitters and batteries. This was what rode in the nose of the rocket as it climbed, made the now familiar shift in direction to take a southeasterly tack, and continuously picked up speed. As he watched from outside a communications trailer, Stehling heard a col-league who was inside on the phone with John Hagen in Washington shout a question to him about the rocket's flight.

Stehling said the rocket seemed to be performing as designed. The lick of flame from the motor was growing smaller as it flew farther from view and then passed the point where TV-3 Back-up had broken apart. The contrails behind the rockets got wiggly as it flew through layers of the upper atmosphere, but the flight was stable.

A few hundred yards away from Stehling, the blockhouse door swung open and a dozen or so men burst out, their eyes trained on the sky. They stood and gazed. A tinny loudspeaker broadcast a stream of in-formation based on radio and radar reports.

The first-stage rocket, having exhausted its fuel, died.

The second stage separated and fired.

The nose cone protecting the third stage fell away.

The second stage died and began to coast.

The third stage began to spin, as required, to ensure stable flight. Then it fired, separated from the second stage, and burned just long enough to reach the edge of space, where the ball would be released.

On the ground Kurt Stehling had the strange feeling that everything had gone too well. The second stage, for example, had never flown before, and yet it performed perfectly. After so many tries, when so many things went wrong, it was hard to accept this success. Others had similar feelings as they came together at Hangar S to await word from a San Diego tracking station that could confirm that the ball was indeed orbit-ing the Earth.

A few miles south, at Patrick Air Force Base, radar expert John Neilon dismissed the military crew that manned the station as it tracked the rocket's departure from the Florida coast. "They weren't really part of Vanguard and just wanted to go home," he would recall. He ran to his car and then drove north on A1A to the Cape. After he was cleared through the gate by guards, Neilon drove quickly to the hangar. He wanted to be there if San Diego reported receiving signals from the satellite.

The men Neilon joined were exhausted, excited, nervous, hot, sweaty, and thirsty. (Vending machines that dispensed sodas and juice were broken and the water at the Cape tasted so bad no one would drink it.) They anxiously reassured each other with comments about the rocket's performance and the velocity of the final stage—almost five miles per second—which was more than enough to put the payload into space. Amid the buzz, one Martin Company man raised his voice and said, for all to hear, "If this bastard has gone around, we're gonna have the biggest party ever seen on the Cape."

Word of the satellite's signal, if it was heard in San Diego, would be flashed immediately to one of the upstairs offices in the hangar. The room where the Teletype was located was filled to overflowing and men hovered over the machines. A big coffeepot boiled over, and no one seemed to care.

After a ninety-minute wait, Stehling's nervous stomach and even more nervous mind drove him out of the room and down the stairs. He was beginning to think about all the things that could go wrong. Then, as John Neilon would recall, Daniel Mazur bolted out of the office to the balcony and shouted that San Diego had received Vanguard's signal.

"They've got it on the yagies!" he said, using techno-slang for the type of equipment that had captured the sound.

Stehling was caught on the stairs as overjoyed men poured onto the balcony shouting and whooping and throwing all kinds of loose objects, even their shoes, into the air and down onto the crowd below. As Stehling turned to rush back upstairs, he ran into a young woman, a secretary, who had started down. They grabbed each other, to keep from falling, and for a moment seemed to be dancing.

Soon after the initial excitement and celebration began to subside, planning for the party to celebrate the bastard's flight began in earnest. The Martin Company men, whose firm would pay the tab, picked the fancy Trade Winds, General Medaris's haunt down in Indialantic. Enough liquor flowed that more than one top Vanguard official was tossed into the hotel pool. John Neilon would remember it as one of the best nights of his life.

★

The Vanguard satellite wasn't designed to do anything more than the first *Sputnik* had done. It was a radio beacon, and any scientific advances derived from its flight would be based on observations made from the ground. Still, Vanguard's success meant that America had matched the Soviets in the number of satellites put into orbit, and for the men behind the rocket it was a vindication of their design, management skill, and craftsmanship. Counting the last-stage rocket, which weighed fifty pounds, they had managed to use a smaller, less expensive vehicle to put more total weight into orbit than Juno. And they had established it in an orbit that was much higher. *Explorer* flew so close to Earth that it would come down within a few years. Vanguard was so high it could stay up for decades, perhaps centuries. (Vanguard was still sailing along, undisturbed, as of 2007.)[7]

But as impressive as the orbit was, Vanguard did not bring the United States close to the USSR when it came to firepower and scientific achievement. With their two satellites the Soviets had a forty-to-one advantage in total weight put into space. Far more importantly, they had managed to launch Laika—a mammal with humanlike biological requirements—into orbit without killing her right away.

Laika's feat related directly to the most romantic goals envisioned by politicians, experts, and laypeople in both countries: putting men, and perhaps women, in space. More than satellites, the concept of manned missions suggested a new age of exploration, similar to that of Columbus, when voyages into the unknown changed the shape of the world as known to Europeans. As with Columbus, these great adventures would have obvious military implications, what with space being the ultimate high ground. But in escaping the Earth, men would also claim powers and experiences once reserved for gods.

By the early 1950s, philosophers, historians, and social commentators had already begun to consider the deep effects that the emerging scientific age would have on humanity when the space race began. *Sputnik* accelerated this process, and soon after it was launched, Charles

Frankel of Columbia University speculated that much that was essential to human experience was about to be challenged. Time and space could be conquered with travel at previously unimagined speeds, and natural phenomena as small as the atom and as big as the weather could be subject to manipulation. Humanity would have to find new ways to rule science and technology, he warned, in order to preserve the good in society and avoid being tyrannized by change.

These high-minded concerns would become the focus of articles, books, and debates in the years to come, and many of the changes Frankel predicted, including a broad conflict between science and religious conservatism, would come to pass. (Campaigns against the teaching of evolution would reflect the extreme right's discomfort with science, while anti-nuclear activism would be generated mostly on the left.) But in the short term, the two nations leading the way into the new era would focus on the tactical aspects of the military opportunities at hand and the technical challenge of opening the heavens to human beings.

In military circles, space was imagined as a place for global surveillance, orbiting weapons, stations manned by soldiers, and the transport of troops. Of course, the offensive uses of space would have to be countered, so the Pentagon also sought the development of antimissile systems, a dream that it promised would come true soon. On both the Soviet and American sides, defense experts warned about the threat posed by the enemy as they argued for bigger budgets. In late winter 1958, the Soviets would even raise the hysterical complaint that the Americans intended to create military bases on the Moon.[8]

But as vital as weapons projects seemed to the Cold Warriors, they were required to conduct much of their work in secret. In contrast, the work done to advance the chances of putting human beings in space was widely publicized and the subject of intense public interest. In early 1958, Dr. John Stapp, the rocket sled rider, now a colonel, revealed the positive results of space travel experiments he had done with chimpanzees. Anesthetized, strapped down, and rocketed along a track, the chimps had been subjected to much more stress and pressure than any space flier might experience and survived without injury. This augured well for the men who would put themselves atop rockets.

Weeks after Stapp reported on his work with lesser primates, a six-foot-tall, twenty-three-year-old Air Force enlisted man named Donald Farrell climbed into a model of a space "cabinet" for seven days of isolation. Six feet long, three feet wide, and three to five feet high, the capsule was a tight fit for Farrell, whose physical and psychological reactions to the loneliness, artificial light, and low air pressure, which would simulate conditions in orbit, were to be studied. The experiment would also give scientists the opportunity to test equipment for recycling air, delivering food, and handling human waste in space.

Game from the start, Farrell predicted his friends at Randolph Air Force Base in San Antonio would be ready "with a fire hose" to bathe him after seven days without washing. He was sanguine about giving up smoking, and eager to help "conquer space." During his "flight" Farrell would throw switches and perform other small tasks as commanded by orders flashed on a small TV screen. For entertainment he picked some recordings—Gershwin and Cole Porter—that were piped into the cabin, and he was allowed to read a bit.

For a week the sedentary adventure of Airman Farrell, pretend space traveler, was played out in the press. On Wednesday his superiors expressed relief that their man didn't like rock 'n' roll because he was requesting music all the time and they would have had to listen along. On Friday one of the physicians for the project predicted Farrell would gain some weight during his mission because he was "a real chow hound."

In fact, Farrell lost weight—about four pounds—but otherwise fared well. He never felt any fear. He seemed unaffected by the isolation and adapted readily to the artificial cycle of work, rest, and sleep imposed by Air Force scientists in lieu of normal daylight and darkness. "Really and truly I could like to make a trip to the Moon," he said upon emerging. After some time in an Air Force hospital he went home to the Bronx for a hero's welcome from the borough president, and then appeared as guest of honor at the Jet-Age Conference of the Air Force Association, which was held at a Manhattan hotel.

While no one questioned the success of Farrell's "mission," experts in the study of men in simulated space flight understood that the experiment had provided him with more diversion, amusement, and human contact via intercom than a space traveler could expect. As a result, the flight of Donald Farrell had little value beyond boosting the country's morale.

The nation would undoubtedly have been less optimistic had the authorities reported at length on the results of more rigorous studies done in the same era. These involved men who were sealed in much smaller capsules and were required to perform more tasks under extreme isolation. Under the stress, hallucinations were common. Some heard sounds that weren't there and others saw faces on the dials of the simulated capsule instruments. One flier was sure a large hole had opened in the floor of his cabin. Another felt as if his limbs were becoming swollen and distorted. A third became so certain that a TV tube on his console was on fire—it wasn't—that he had to be let out. In each case these problems, which were blamed on sensory deprivation, arose within the first day and a half of a simulated spaceflight. They pointed to a problem that would become fodder for science fiction—the prospect of going mad in space.

The spaceflight simulations done on Earth, and in high-altitude balloons, were adequate to test for some of the physical and psychological effects of spaceflight, but they could not address every concern. The effects of weightlessness, for example, could not be studied thoroughly on Earth and no one knew for certain the kinds of radiation hazards a space traveler might encounter.[9]

Answers to some of these questions were supposed to come from the *Explorer* satellite and the scientific instruments from James Van Allen's group. The satellite was designed to broadcast the number of particles hitting its Geiger counter to the network of receiving stations that had been established around the globe for the IGY. But because information was available only when the satellite was directly over a station, the results were spotty.

The early data generally confirmed reports on low levels of cosmic rays suggested by Van Allen's many rockoon experiments in various regions of the atmosphere. But there were times when the satellite passed over a station and the count broadcast down to Earth suddenly dropped to zero. This was the kind of result Van Allen did not expect and it was a puzzle. Equipment failure was one possibility, but the devices put into the satellite had been thoroughly tested, abused really, before they were installed, and had proved capable of surviving very rough treatment and harsh conditions.

Van Allen hoped the mystery of the zero counts would be resolved when more equipment was launched in *Explorer II*, but the failure of that rocket to go into orbit ruined that experiment. Then, on March 26, a Jupiter barely muscled *Explorer III* into a lopsided orbit with an apogee (highest point) of about 1,700 miles and a perigee (lowest point) of 110 miles. The perigee was so close that the Earth's atmosphere and gravity would pull it down in a matter of weeks, if not days.[10]

The orbit was all the more frustrating because *Explorer III* carried a much improved payload. The key addition was a tiny recorder and player, assembled by George Ludwig, that would store readings from the Geiger counter and then, in response to a radio command from the ground, broadcast the information to a tracking station in concentrated bursts. With this new method, two hours' worth of data could be transmitted in six seconds. Scientists would be able to receive counts on cosmic ray "hits" recorded by the satellite for an entire orbit.

The large volume of information was funneled to Washington, D.C., where a computing center processed it. Van Allen was in town when the first report on readings gathered from an entire 104-minute orbit was finished and he took a taxi to pick it up. He went back to his hotel room with a printout of the data and, using his trusty slide rule and graph paper purchased at a drugstore, began plotting out the numbers. The results were similar to those from the first satellite. Low levels of cosmic rays were found close to Earth, but the number of hits recorded by the Geiger counter rose as the orbit got higher until they suddenly plunged to zero. Minutes later they would rise again and then fall. Gradually Van Allen concluded that two satellites wouldn't have the same instrument problems. By 3 A.M., as he went to bed, he was certain some natural force was at work.

While Van Allen had been working his numbers, his assistants in Iowa continued to mull the problem of the data spikes. One of them, Carl McIlwaine, had just returned from the Arctic, where he had been launching small rockets. (His main scientific interest was polar auroras, and he had been firing directly into their glowing light.) Being new to the problem of the *Explorer* data, McIlwaine had a fresh perspective.

What if the problem wasn't too little radiation, he wondered, but too much? Under those conditions the Geiger counter might be unable to count the impact of all the particles and, instead, report nothing.

The answer came when McIlwaine subjected the same kind of Geiger counter used in the satellites to the energy produced by an X-ray machine. As the power was turned up, the counter eventually went on the fritz and indicated a count of zero. *Explorer* wasn't passing through a zone of no cosmic rays, but rather through a region of space where radiation levels were extremely high, he concluded.

Before it succumbed to gravity and atmospheric friction, *Explorer III* would provide useful data for forty-four days. In this time Van Allen and his colleagues would determine that they had discovered, at an altitude of about 600 miles, a band of energized particles held in place by the Earth's magnetic field. The particles moved between the poles and drifted around the planet, and they existed in these high concentrations to an altitude of perhaps 8,000 miles. Their energy was so intense that an astronaut exposed for ninety minutes would receive a week's worth of the highest radiation dose deemed safe.

With the information from *Explorer II* and the riddle of the zero counts solved, James Van Allen and his students had made the first scientific discovery based on a probe of outer space. He presented their findings at a conference at the National Academy of Sciences. It was there that someone first described the discovery as a radiation *belt*. Months later the term "Van Allen Belt" would come into use. Although the belt presented an obvious hazard to any living thing in space, Van Allen estimated that 100 pounds of lead shielding would protect a man in orbit.[11]

This weight would add to the challenge of actually putting a man in space with an existing American rocket. By spring of 1958, the idea that space travel would begin with an astronaut in a protected "capsule" thrown by a rocket into a simple orbit had surpassed other concepts, including that of a space plane. Considering all the other equipment required to protect him, and the weight of the fellow himself, a two- to four-ton capsule was envisioned as the minimum vehicle required

Within the community of contractors, scientists, and officials who would develop and pursue America's space agenda, manned spaceflight was the most glamorous goal. This achievement would be the next point of competition with the Soviet Union, and it required substantial rocket power. For the moment, nothing in the American arsenal was capable of launching a capsule as heavy as the one imagined by designers. In contrast, the USSR's R-7, having already sent Laika on a trip into space, was more than adequate for the job.

Just in case their American adversaries had forgotten who was the weightlifting champ, on May 15 the Soviets used another R-7 to hurl what seemed like a super-*Sputnik*—three tons, including more than 2,000 pounds of scientific instruments—into an orbit that reached almost 1,200 miles high. This satellite was actually the kind of payload that the Chief Designer had first expected to launch as *Sputnik I.* Included were instruments to measure radioactivity in near space and a recorder-playback device similar to the one Ludwig had made for *Explorer II.* Unfortunately for the Soviets, their recorder went up without proper testing and promptly failed.

In time, the world would learn that the Soviets had actually experienced a much bigger failure two weeks prior, when an R-7 rocket broke up less than two minutes into its flight. But in the short term the accident was kept secret. The world knew only that the Soviets had a new satellite, which allowed Khrushchev a chance to compare his spacecraft to America's "oranges" and suggest that the United States would have to put many bushels' worth of *Explorers* into space to match the *Sputniks*. The size of the payload would remind American citizens that the space race had chilling, Cold War implications. *Sputnik III* was so heavy that its orbit proved that the USSR could send one of its crude atomic weapons to devastate an American city. On the day that the big satellite flew, Wernher von Braun said the United States would need eighteen months to match such a feat.

★

After so much space news and the reassurance provided by the *Explorers,* big *Sputnik III* did not cause public uproar in America. In general, the nation seemed to be coming to accept the space competition. President Eisenhower had announced the creation of the National Aeronautics and Space Agency (the word "Agency" was later changed to

"Administration") and Congress was setting both a long-term agenda for space and establishing the apparatus to pursue it. Under this plan, most science and exploration would be done by the civilian NASA while the armed forces would continue military work, from surveillance satellites to ICBMs. Of course these duties were still subject to fierce interservice competition, with the Air Force and the Army competing to control land-based missiles.

In the political world, few found it necessary to respond with any vigor to *Sputnik III*. An exception was Senator Stuart Symington, who called the latest Soviet success "another grim warning for all but the blind to see." He accused Eisenhower of following a dangerous policy of "drift and dream" and demanded "an accounting to the American people." This attack, when others were silent, may have had more to do with the former Air Force secretary's concern about defense dollars and where they might be sent than satellites. Whatever the motivation, it led to nothing.[12]

NINE

Opportunists and Adventurers

The first satellite.

The first animal in orbit.

The first American satellite.

The first scientific discovery in space.

Each *first* in the rocket/missile/satellite competition prompted a hullabaloo. Subsequent incremental achievements were noted, but never caused the same excitement. Indeed, after just six months—the time required for all of these firsts to occur—the world seemed to adapt to space exploration. The presence of something man-made in the heavens was becoming unremarkable. The Japanese and the British reported successful rocket launching that scared no one, and diplomats began to consider the need for rules, perhaps even treaties, to promote normal relations in this new realm.

Ordinary citizens in the two countries that had gone to space could set aside their fears and celebrations. Americans saw that *Sputniks* had flown and fallen to Earth, causing no harm. Their presence had become so benign that many newspapers published daily reports—similar to weather forecasts—that offered information on the travels of the satellites across the night sky and advice for viewing. As calm returned, UFO sightings in the United States dropped by about 75 percent. In the

Soviet Union, officials began spreading a new national motto—Catch UP!—as the country refocused on its long-running effort to match the West's standard of living. Satellites were fun, but the nation also needed better tractors, new housing projects, and more wheat.

Americans needed to focus on economic goals, too. A recession, which arrived at about the same time as the first *Sputnik*, had brought record-high food prices and contributed to a sharp rise in unemployment. As a result, people had plenty of mundane concerns—the price of hamburger, job security—and, in general, did not dwell on either the practical aspects of the space competition or the deeper meaning of humanity's advances in new technologies and new realms.

The exceptions to the life-goes-on rule could be found in communities where new technologies were created and deployed and in certain professions, most notably in the arts and the media, where the quest for understanding could be indulged and even promoted. Three different space-related Broadway musicals were written in the months after the first satellite went into orbit. In the same period Rod Serling developed the pilot for his television program *The Twilight Zone*, which would become a highly regarded series and later a classic of Cold War camp. Just as play helps a child gain mastery over certain psychological concepts, this television series would signal America's effort to adapt to challenging concepts such as nuclear destruction and space travel.[1]

While artists and writers dealt with space as an intellectual conceit, the men and women in the rocket business were busy constructing new industries, communities, and scientific disciplines. Based on expensive and esoteric technologies, these endeavors attracted highly competitive and adventurous people, the kind who could imagine themselves at the center of something exciting. Most readily adapted to a life in which potentially deadly experiments were normal, everyday events. A few were so enthralled that they couldn't bear being away from it.

Young Jay Barbree, the lanky radio reporter with the Georgia drawl, felt the pull to be around this space business from the moment he heard about *Sputnik*. The yearning grew more intense after he stood outside the gates to Cape Canaveral for the TV-3 fiasco and, later, for *Explorer*'s success. Here was something big, an enterprise found in just a few places in all the world, and he could be part of it even though he wasn't an engineer, a scientist, or a technician. He didn't have to be any of these things to get close to the action. He could do it as Mobile Mike.

Mobile Mike flitted up and down the main roadways in the Cape area, in a tiny red car topped by a model rocket that was poised to blast off its roof. Mike's odd vehicle, an Italian import called an Isetta, also seemed to be missing a rear wheel. In fact there were two small closely spaced wheels in the rear,which gave the car the look of a tricycle.

Styled like the bubble-shaped space car that George Jetson would fly through a cartoon universe, the Isetta that Barbree drove so he could deliver traffic reports as Mobile Mike was an attention-grabbing hit for WEZY. In the mornings and evening he would cruise A1A, U.S. Route 1, and other roads leading to and from the Cape and then call in his reports to various programs. Always, he was the motorist's friend.

"You might as well pull over and get yourself a cool one because you ain't goin' nowhere," he'd say during afternoon jam-ups. When the Army Corps of Engineers raised a drawbridge at the wrong moment, he'd complain with enough ferocity to satisfy all of his aggravated listeners.

Mike's audience grew larger every day as the number of people employed at the Cape rose from 14,500 to over 18,000 in his first year. Pressure on local housing and other resources increased. In May 1958 alone, a thousand new homes and apartments would become available, but this wouldn't come close to satisfying the demand. For this reason, some commuters started their day forty miles west, in Orlando, and prayed that Mobile Mike would guide them to the Cape's gates.

Occasionally, a lucky worker on his way to the test site or a space migrant just pulling into town would find himself (or herself) on a causeway eastbound over the Banana River when a missile was fired from one of the seaside pads. Thomas O'Malley had this good fortune on St. Patrick's Day as he drove his Volkswagen on the last leg of his move to the Cape from New Jersey and saw the first successful Vanguard blastoff. O'Malley, who would eventually press the button that sent John Glenn into orbit, was thrilled to be joining the space effort at the Cape. But as a family man, he worried for the months it took him to find a proper home for his wife and children. He also felt a twinge of guilt about moving them from a comfortable suburb outside New York City to a comparative wilderness. "For the longest time we had to drive to Melbourne for a real food store and to Orlando for clothes," he would recall.

Single men like Jay Barbree didn't suffer as much over creature comforts. Most could be content with crash pad motel suites or rooms they rented from Cape families. Barbree, who was so busy that he barely had time to sleep, found a little apartment on 4th Street South in Cocoa Beach. This became his base for a life that was suddenly full of excitement. Ever ambitious, he wrote a one-paragraph letter to an executive at the National Broadcasting Company in New York, offering his services. He soon became a space correspondent for both NBC radio, at $35 per report, and the NBC television network, which paid him $50 every time he sent some news. The job gave him the credentials for covering the big stories at the Cape and, on a busy day, he earned more than he could make in a month as Mobile Mike.

Business was generally good for Barbree because it didn't matter to the network bosses whether a rocket flew or failed. They just wanted to be on top of the story. In fact, sometimes crashes made for better copy, and in the summer Barbree would report on the explosion of an Air Force rocket intended to orbit the Moon that didn't fly ten miles and a Navy Polaris missile that blew up so soon after launch that one big chunk almost hit the launch pad, and other flaming debris, including some pieces large enough to make explosive sounds at impact, fell into the Banana River near a trailer park. (The river runs south from the space center, separating Merritt Island from Cocoa Beach.) In Barbree's memory, at least one trailer park resident was so frightened she came running outside covered only by a towel.

The Polaris, which was purposely destroyed because it went dangerously off-course, was an intermediate range ballistic missile, IRBM, with a range estimated at 1,500 miles. Days after this incident, Barbree and prankster friend went out in the dark of night to put up a sign on the road alongside the river. Looking something like a roadside marker at a historic site, the sign established a certain spot in the stream as the resting place of an IBRM, or In the Banana River Missile.

It's remarkable that Barbree had the time or energy for such elaborate jokes, given that he continued to work as Mobile Mike. Mike was a popular local celebrity, and in addition to his traffic assignment he was asked to perform certain duties befitting his position. One called for him to be part of the Fourth of July parade in the city of Cocoa. He drove a big open car, donated for the day by a local dealership, that carried a beautiful contestant in the Miss Brevard County contest to the park where the competition would be held.

At the pageant, where young women in one-piece bathing suits paraded on a decorated flatbed truck, Barbree took special notice of a dark-haired contestant—a former Miss Lockheed Aircraft—who was sponsored by the Red Rooster Restaurant. Called to walk and turn in front of the crowd, Jo Ann Reisinger did her best to appear relaxed and graceful. Having entered the contest to please her mother, who thought it would help her overcome shyness, Jo Ann was shocked when she won.

The victory made her eligible for the Miss Space pageant, which was held soon after. With the help of her parents Jo Ann put together an outfit of shoes painted gold, red tights, a gold vest, and red leotard. She topped it off with a hat that looked like a globe and twin antennae tipped with blinking lights. With a toy ray gun as an accessory, she won again. For a moment she was, perhaps, as much a local celebrity as Barbree. She went to his apartment for a party. Despite her mother's worries about their age difference (she was eighteen, he was twenty-four), they would begin a relationship that would lead to a lasting marriage that was still going strong fifty years later.[2]

★

When Mobile Mike met Miss Space, the frenzied effort that produced first the Vanguard collapse and then the *Explorer* success was giving way to the more deliberate pursuit of scientific progress as the government prepared to open its civilian space agency. Having urged the press to remember that their program was still experimental, the Vanguard group surprised no one with four failed satellite attempts between March and September of 1958.

In the meantime the Army missile group maintained its status with additional successful tests, including the first recovery of a nose cone from an intermediate range missile. The Army group's leaders seemed to be building a real institutional presence that anticipated a long-term future. In the summer of 1958 they initiated the first youth intern program at the Cape and brought two recent high school graduates to the missile range to serve as aides to space engineers. One was a young man from Huntsville named V. L. Pinson Jr. The other was La Verne Hardin's daughter, JoAnn.

From early June until she left for college in September, JoAnn Hardin would help program managers by doing a variety of chores. She first met General Medaris when he burst into one of those offices demanding, "Get me Washington on the phone!" A startled JoAnn dashed to a communications room for help and later learned not to fear the general. He was only "barking smartly" as military people said, and he did this quite often.

Everyone at the Cape knew Medaris's habits because at this time it was still a fairly small place and space work was like a fraternity. JoAnn would form close attachments to members of the Army group, who encouraged her to make space a career. Kurt Debus, the chief of missile firing, treated her like a daughter. When she mentioned studying history at college he said, "Don't worry about that history. You're going to make history yourself." Other supervisors, including Austrian Karl Sendler, a German missile veteran, made sure she was more than a helper, giving her meaningful work to perform right from the beginning.

Just days after she started, JoAnn was assigned to operate a small instrument mounted on a tripod, in order to track a Jupiter A fired from Launch Complex 6. The night launch lit up the Cape and JoAnn carefully followed the rocket into the sky to record a ring of flares that fired as the first and second stages separated. The successful flight was followed later by a launch that ended in disaster with a rocket exploding and JoAnn racing outside to gaze in awe at the beautiful, multicolored bits of flaming wreckage that flew high and then tumbled down to the ground.

In those days, before stricter regulation, people tended to look up rather than run for cover whenever they heard or felt an explosion. Since rockets were aimed to fly out over the ocean, this practice was relatively safe, and the only injury JoAnn would recall involved an excited fellow who had climbed onto a roof to get a better look and fell off. Dr. Debus wrote a stern stay-off-the-roof memo after this mishap, but one could hardly blame the guy for seeking a better view of a launch. The excitement at the Cape was intoxicating, so much so that when her internship ran out, JoAnn would be reluctant to depart for college. She thought, for a moment, about just taking a job instead. Some of the kids who graduated high school with her did forgo education and earned good pay doing interesting work in the space industry. One was red-haired Dalton Cairnes, who married a local cheerleader and made a life as a Cape worker.

As a member of the ABMA family who had earned the notice and affection of so many professionals, Hardin didn't have the option of taking a low-level job. Instead they taught her the habits of scientists, including making a daily record of her work, counseled her on her academic program, and encouraged her to consider bigger ambitions. She got more inspiration in July when *Explorer IV* went into orbit with instruments that confirmed the existence of Van Allen's radiation belts. The data gathered on radioactivity in space helped to answer some questions about the conditions men would encounter during spaceflight.

★

Despite the success of the *Explorer* satellites, the United States still lagged behind the Soviets when it came to studying how rocket flight might affect a living creature. America lacked the firepower to send a large animal like Laika aloft. But an experiment with a smaller animal was possible and so the Air Force began sending mice into space. The first, named Mia, was launched in the nose cone of a Thor-Able rocket—a modified Thor—that was never recovered. The second—Mia 2—suffered the same fate, sinking in the sea.[3]

Finally, on an otherwise slow July day at the Cape, Air Force officials succumbed to pressure from curious reporters and introduced them to their third experimental subject before she took her ride. Arriving at the press conference in a big glass jar, the three-month-old white mouse was presented by Air Force Colonel Charles Mathison and Captain C. E. Griffith, who was in charge of everything related to the mouse.

With Griffith doing most of the talking, and Mathison handling the mouse, they explained the life support system that would provide oxygen, water via a wick, food in compressed pellets, and air-conditioning to last six weeks. Her house, an ingenious contraption, would cradle its resident in a mesh screen that would swivel in all directions so she would always feel secure on her feet. A sensor attached to her chest would record the mouse's heart rate, which would be radioed to the Cape. Once her nose cone splashed down in the Atlantic it would be kept afloat by a balloon. Searchers would be guided by a radio beacon and a strobe light. The mouse would be protected from sharks by a chemical repellent. Upon recovery, detailed exams would yield more information on the effects of space travel.

For a press corps accustomed to terse briefings by engineers, military men, and scientists who offered only the barest information about their mission, the display of all the equipment and the furry little experimental subject was a treat. The mouse, with her tail marked in stripes of pale pink and green (for identifying purposes), was in a playful mood, ignoring the cameras and exploring the jar with her twitching nose. The people around her caught the spirit. Questions about the mouse's name led to nominations. AP correspondent Jack King, a recent transfer from Boston who was quick with a joke, suggested "Wickie Mouse."

King had good arguments to support his proposal. Obviously Wickie Mouse would catch the public's ear because it echoed Disney's most famous son, Mickey. King also noted that the mouse seemed to enjoy drinking from the wick that delivered water in its jar. And finally, the name would honor the youngest reporter in the crowd, who was also the object of much flirting, Wickie Stivers. The name was adopted by acclamation.

As soon as the news about Wickie Mouse went out, the young woman who lent her name was deluged with calls from radio and television stations requesting interviews. Her bosses at Reuters in New York, whom she served as a stringer, congratulated her with cheers of "Good show, old girl."

Wickie Stivers's hope for her namesake's safe return from the edge of space had to be tempered with her knowledge of the risks involved in every rocket launch. On the day before Wickie Mouse met the press, the most powerful American rocket—an Air Force Atlas-B—had failed in especially fantastic fashion. The event was reported in vivid detail by Milton Bracker of *The New York Times*, who noted that a halt was called when the countdown reached three seconds, but the first stage ignited anyway, sending the Atlas- B wobbling skyward with the motion of "a hooked game fish." Moments later, "the tortured monster was swallowed up by another puff of smoke that looked like the mushroom top of an atomic explosion." Debris, including some pieces large enough to make loud crashing sounds on impact, fell on both land and sea.

The obvious danger in riding rockets would move even the crustiest reporter to regard Wickie Mouse as a brave little creature *if* she had any idea what was in store. But of course she didn't suspect a thing as her multimillion-dollar vehicle blasted off from the Cape just after 6 P.M. on Wednesday, July 22. Traveling at speeds up to 15,000 miles per hour, Wickie's nose cone was expected to land 6,000 miles to the southeast, near Ascension Island. Ships and airplanes were poised to search for the mouse house and recover it.

Although it was equipped with a strobe light and a flotation colored fluorescent orange, in the vast ocean it would be difficult to find the nose cone without the aid of its radio signal, which failed to broadcast. One aircrew did spot the light in the water, and dropped a beacon to mark its place. However, they had to depart when engine trouble arose. When others arrived at the same coordinates they found nothing. After forty-eight hours, and a great expenditure of manpower and fuel, the mouse hunt was called off. Officials admitted that while Wickie's loss was unfortunate, they were more troubled by the fact that they had failed, for a third time, to retrieve an object sent into space.

The news was bad for the mouse, but Wickie Stivers's fortunes rose even as the nose cone sank. She handed out hundreds of "Wickie Mouse Club" cards, which she had printed, and this kept the story going. But she probably didn't need this promotion. Television producers who noted that she was pretty, and accustomed to being in the public eye, invited her to Manhattan for a week-long visit so she could appear on three different game shows: *What's My Line*, *I've Got a Secret*, and *To Tell the Truth*. The flight to New York was her first trip in an airplane and would inspire her eventually to leave Florida, become a flight attendant, and travel the world. For Wickie Stivers, who just happened to be a budding reporter in Cocoa Beach when it all began, the space race led to a new life. For others, the first summer of the space race brought brief but intense experiences that would stand out in their minds for decades to follow.[4]

★

Radio specialist Bradford Whipple's first brush with *Sputnik* had come during its initial orbit when he heard its beep-beep signal at an Air Force listening post in West Germany. He had joined an airman called Hogjaw to track the signal and found himself a bit angered and resentful about the USSR's achievement. He did not expect another run-in with the Soviet space program on August 8, 1958, when he drove out of Darmstadt with a beautiful young German woman named Anke Hansen, bound for Brussels and Expo '58.

The first world's fair of the postwar era, Expo '58 was enormously popular and would draw 51 million visitors. The crowds drove up prices for food, hotels, and everything else tourists needed. For Whipple the prices added extra stress to a trip that started badly when Anke's mother made it clear she disapproved of the whole idea. But after a tense journey, the couple's spirits were lifted when they spotted the large and futuristic buildings of the fair in the distance. Like everyone else who attended, Anke and Brad were impressed by the sight of the 335-foot tall Atomium, a building and sculpture that towered over the sprawling fairgrounds. Fashioned after a microscopic iron crystal, it was made of nine shiny metal spheres connected by tubes. The spheres were large—sixty feet across—and the tubes carried people in escalators. From afar, it looked like a space station built out of giant Tinkertoy parts, ready for habitation.

After paying their 60 cents admission, Anke and Whipple could indulge in carnival rides, lounge at a full-scale Bavarian brewery, or tour scores of pavilions devoted to various industries (aluminum, mass media), corporations (Bell Telephone, Coca-Cola), and nations. To suggest visions of the future, many of the exhibition spaces were built with domes, shiny metal beams, and great expanses of glass. Others were designed with dramatically curving surfaces, cantilever roofs, and extensions that seemed to hover in the air.

The Soviet pavilion was a huge glass and steel box, about seven stories tall, nearly 500 feet long and half as wide. The structure covered a single, open floor filled with exhibits. In an effort to demonstrate the superiority of their political system, the Soviets spent a reported $50 million on their exhibit (four times the amount the United States invested in its pavilion) and filled it with the best goods, inventions, and even people they could find, including the Bolshoi Ballet.

As they climbed the grand stairway and entered the Soviet hall, Whipple and Anke were met by a statue of Lenin dressed in a great long coat. Standing on a pedestal that was at least twelve feet tall, the statue rose to within feet of the sixty-foot-plus ceiling. Around him were airplanes suspended from arched metal supports and displays of Soviet products, including cars, heavy machinery, and the biggest but flimsiest-looking tractors and trucks ("terrible welds") that Whipple had ever seen. Inexpensive Soviet goods, including Laika cigarettes with a picture of the space dog on the pack, were popular with fairgoers. Whipple bought some of the cigarettes and a Soviet Pilot watch — "guaranteed for one year against all defects in materials and workmanship" — that would break in two months. He also spent some time staring at the most crowded attractions in the pavilion: two model *Sputnik*s suspended from the ceiling, near the statue of Lenin.

The sight of so many Europeans gathered around the satellites galled Brad Whipple, but he was even more upset when he considered how America's hall and exhibits compared with the Soviet pavilion. The United States had approached the fair in a more creative way, building a circular structure out of glass and gold-colored metal columns that shone warmly in the sunlight. The result was a much more welcoming structure but one that also might seem, to some eyes, wimpy in comparison with the Soviet pavilion.

The warmer-and-fuzzier approach continued inside the American hall, where the exhibits were more artful and provocative than boastful. Similarly, the people sent to represent the United States at the fair were urged to be modest so they might avoid the "ugly American" stereotype. At least one of the displays in the building went beyond modesty to explore, through photos and film clips, America's social problems, from racism to poverty to juvenile delinquency. The idea for this display came from an advisory group convened at the Massachusetts Institute of Technology and led by economist Walt Rostow. It was called "American Idealism in Action" and was meant to demonstrate the gradual progress of democracy and how an open society deals with its challenges.

The subtlety of the American approach was lost on young Airman Whipple, who was annoyed to see his country's problems on display next to a Soviet exhibit that suggested an all-but-perfect society. The experience didn't help his mood, which was bad already due to lack of sleep, the effects of fair food, and Anke's uneasiness. Her mother's complaints about Brad and their shared vacation still hung over the young couple and they had bickered.

On his third night in Brussels, Whipple went alone to a pub where he hoped to find an escape from the tension in his hotel room. He sat at the bar. Before his third beer he was approached by a man in civilian clothes, with a matching fashionable haircut, who looked at his Air Force–issue shoes and asked, "You Army, or what?" (The shoes were a give-away. Prostitutes around the world also relied on them to identify GIs.)

One question led to many others and soon Whipple and his new friend joined a group of four men at a table. He told them about his posting in Germany and his brush with *Sputnik* and how irritated he felt by the Soviets' efforts to show up America in space and at the exposition. Soon they asked him if he would mind leaving the bar to help them with a small chore.

"It was said like it was nothing serious," recalled Whipple many years later. "I had had four or five beers. I said, 'Sure, why not?'"

Two cars with French license plates waited outside. Whipple got into one with two of the men and they followed the second car through the streets and then down a dark boulevard lined with trees. As they rode, the men explained to Whipple that they were going to break into the Soviet building at the fair and steal a *Sputnik*. They planned to move it to a safe place, take it apart, photograph it, reassemble it, and return it, all before dawn. His job would be to stand watch by the back door of the pavilion.

Still a little tipsy from the beer, Whipple welcomed the chance to do something for his side in the space competition. In the distance the Atomium, lighted and glowing, came into view. The cars stopped along a street that flanked the fairgrounds, where guards patrolled but there were no fences to protect the pavilions. The men got out and silently made their way to the back of the Soviet building.

Whipple watched while one of his new comrades picked the lock and went inside, followed by four others. In a few minutes three of the men returned with the shiny ball. (Light, and just the size of a medicine ball, the model *Sputnik* was easy to handle except for the antennae. They rushed to one of the cars, stowed the satellite, and drove away. Whipple remained outside the pavilion as a lookout for the two Americans who remained inside and he wondered what might happen. He began to have second thoughts about volunteering and with the alcohol wearing off, even began to wonder if his partners in this escapade were actually Americans.

A couple of hours passed. Whipple may have dozed in that time, but he was awake when the men and the satellite returned. They went back into the pavilion and emerged with the entire group. They were excited, like a bunch of fraternity brothers who had just pulled a prank. The men, whom Whipple took to be intelligence officers of some sort, dropped him off at the little hotel, where Anke was asleep. He never saw them again. He would keep the secret of what they had done together for decades.[5]

★

Bradford Whipple and his partners in the Brussels Expo escapade were helped in their effort to keep their secret by the impromptu nature of the mission and its small scale. Since just six men participated, it would be easy to trace a blabbermouth. Another highly sensitive and top secret project conducted in late summer 1958 wouldn't enjoy these advantages. Instead, thousands would be aware of this expensive, high-priority space adventure, even if most of them lacked the proper information to know exactly what had happened.

Planning for the project, which was code-named Argus, had started with the post-*Sputnik* imaginings of elevator-engineer-turned-physicist Nicholas Christofilos. For months he pushed for an experiment involving atom bombs set off in the atmosphere. In this case, he noted, the bombs were not to be considered weapons but merely "a convenient source of a large quantity of electrons."

Christofilos won over so many officials in the lower layers of government that he was invited to Washington for a meeting in a special, locked, windowless conference room in the Executive Office Building next to the White House. There he pitched his theories to the members of the President's Science Advisory Committee, who sat around a huge table flanked by chalkboards. They caught his excitement and, after consulting experts in science and defense, approved the experiment.

As head of the PSAC, James Killian would have to shepherd the proposal through the White House. The job was easier than he might have expected. Instead of having to brief the entire National Security Council, which would have been standard procedure, Killian was asked to go directly to the Oval Office. There in a private meeting the president approved Argus immediately. From that moment, on May 1, to the start of the experiment in September, a huge effort was made to organize Navy ships, Army rockets, and bombs from the Atomic Energy Commission.

Christofilos had theorized that electrons released by an atomic explosion would be caught in the Earth's magnetic field and quickly distributed around the planet until they formed a kind of shell. Scientists like Van Allen would be fascinated to see the effect on auroras and what was called the magnetosphere. More importantly, defense experts wanted to see how all this radioactivity would interact with radar and communications, and whether it could influence incoming ballistic missiles and their atomic payloads. Altogether, the products of these explosions would be called the "Christofilos effect."

Since the rockets would be fired at sea, the U.S. Navy was given primary responsibility. The USS *Norton Sound*, with a crew accustomed to handling sounding rockets, was selected as the launch pad. As the *Norton Sound* steamed toward a position south of the Falkland Islands, eight other ships, including the carrier *Tarawa*, were sent to various parts of the ocean to monitor the experiment. This spot was chosen for scientific reasons, and because its isolation would make it easier to keep Argus secret.

Public sightings of recent atomic weapons tests in the atmosphere over the South Pacific had underscored the difficulty of keeping such a spectacular display out of view. Residents of islands from Samoa to

Hawaii had seen post-explosion auroras and even the flash of detonation from hydrogen bombs code-named Teak and Orange. They had also been affected by radio blackouts that immediately followed the light shows in the upper atmosphere.

While Teak and Orange had been observed by the usual complement of scientists, they were primarily tests of weapons designs. With Argus a much greater effort would be made to gather radiation data from the explosions. Radar and radio listening posts would scan the skies. Small rockets equipped with radiation sensors would be fired into the upper atmosphere and would radio data back to trailers packed with collection equipment. Air Force transport planes loaded with scientific instruments would fly into the radioactive stream. The satellite *Explorer IV*, which had been launched into a special orbit for this purpose, was equipped to report back on the Christofilos effect. And special detectors were constructed out of loops of wire, dozens of miles long, that were laid down on federal lands in Arizona and New Jersey.

On August 27, the *Norton Sound* was stationed about 2,000 miles southwest of Cape Town, South Africa, near Tristan da Cunha, an almost inaccessible group of islands inhabited by a handful of people. The late-winter seas were rough. At about 2:30 A.M., a rocket was fired from the ship's deck, carrying an atomic weapon to an altitude of about 300 miles, where it was detonated. As energy equivalent to between one and two kilotons of TNT was released, a brilliant flash of white was followed by streams of colored light—blue, green, red—resembling polar auroras.

At dozens of sites around the world, teams devoted to gathering data on the blast went into action. One of the most exciting assignments belonged to the crews in Virginia, Florida, and Puerto Rico, who launched the sounding rockets. These were such rudimentary machines that firing required a technician literally to pull a cord and then run as fast as he could to get behind a concrete barrier before the thing ignited. Many of the rockets either failed to fire or went astray in flight. With a successful launch, however, instruments in the rocket's nose would detect radiation as it climbed and fell and radio the data to a physicist posted in a van filled with equipment.

After the first Argus bomb was detonated, the instruments carried by sounding rockets and other monitors found that the energy from the blast traveled more than 5,000 miles through the atmosphere in less than a minute and, as Christofilos predicted, was detected around the world in less than an hour. Some observers likened the effect to a magnetic storm caused by the sun, with the accompanying disruption of radio waves. All those who saw data from *Explorer IV* noticed spikes in its radiation readings as it moved in and out of the "shell" of electrons created by the blast. The effect was charted in many locations, including Iowa City, where streams of data were tabulated by Van Allen's group in an office where a sign on the wall announced, "This job is so secret even I don't know what I'm doing."

Two more atomic bombs were sent into the sky and exploded as part of Project Argus. The last, fired on September 6 from 800 miles south of the first shot, produced the most impressive result: a red-crowned aurora in the North Atlantic. Each of the three explosions created effects that lasted for days. But even the longest-lived effects—caused by Argus III— were gone in a little more than two weeks. This pattern was noticed by scientists in the Soviet Union, where radiation recorders at stations across the great breadth of the country showed sudden spikes in atmospheric activity within seconds of each explosion and then continued detecting the excess energy in the sky.

As expected, the Christofilos effect disrupted some radar and radio signals. But though officials said Argus provided valuable information about possible defenses against incoming missiles, *Explorer IV* moved through the radiation layers hundreds of times with no ill effect. This suggested that the atomic particles would be no umbrella against a rain of speeding warheads.

Project Argus was so complex and ambitious that James Van Allen would call it "one of the greatest experiments in pure science ever conducted." It was, in fact, the first big experiment ever performed in space and the first to produce results that could be observed and recorded at the same time all over the world. But as impressive as it was scientifically, Argus was also a marvel of human organization. Never before had so many people and resources been marshaled so quickly for a massive project that was kept so secret. Six months would pass before the government revealed Argus. This was done only after the Soviets had apparently guessed (or learned from spies) what had happened, and began speculating publicly about the three explosions.

Like so many experiments, Argus produced many more questions than answers, and some scientists suggested it be followed with a more ambitious series of nuclear detonations ascending to an altitude of 15,000 miles. Christofilos himself wrote about the knowledge that could be gained from more explosions, especially if they were conducted during "periods of intense solar activity."

The idea didn't get very far. In October, President Eisenhower would announce a voluntary suspension of atomic testing and the United States began negotiations with the Soviet Union on a formal test ban. When the public eventually learned about Argus, in March of 1959, immediate criticism came from citizens who wondered about the safety of such experiments and the ethics of one country deciding to play around with areas of Earth's atmosphere that no nation ruled.

In the end, the Argus experiment would go down in history as the only intentional space exploration program ever conducted with atomic bombs. In time, Herbert York, who was deeply involved in Argus as the Pentagon's chief research scientist, would regard the mere fact that it was conducted at all with wonder and even amazement.

"Christofilos was difficult to deal with, the kind of fellow who wouldn't take no for answer," York would recall decades later. "He wouldn't even take 'slow down' for an answer. And his ideas were always expensive. But at that time the possibilities were wide open. We were investigating every possibility and I first gave the green light to doing it. The idea went up the chain of command from there."

Cooler heads, budget restraints, and international concern about radioactive fallout soon ended the opportunity for projects like Argus. York would look back on Argus as the type of experiment that could only have happened in *Sputnik*'s wake and with a personality as strong as the Greek's behind it. "It was interesting science, but the chance that it might create an antimissile shield in the sky was always small," added York. "Imagine proposing such a thing today. You'd never get away with it. But in that one moment, you could."

In fact, several outlandish research proposals made just a few months after Christofilos floated the Argus idea were never pursued to completion. Hydrogen bombs were *not* employed to break ice at the North Pole and aluminum balloons were *not* sent to float around the moon. Perhaps the most outlandish of the unrealized dreams, dubbed Project Orion, proposed that a huge ship—4,000 tons, to be exact—be powered into space by a series of controlled nuclear explosions.

Orion received modest research funds because it had some merit in the eyes of many experts, who thought its power system solved many of the problems posed by extended space missions. One flight alone, it was theorized, could carry enough matériel and personnel to establish a Moon base. The group behind Orion, which included the physicist Freeman Dyson, was so excited by its potential that they adopted the motto "Saturn by 1970." Of course, Saturn was not on the national space agenda, and Herbert York saw too many problems to merit a substantial investment of federal dollars. Soon the cost, combined with nuclear test bans and public concerns about nuclear technologies, made the proposal untenable.[6]

TEN

Eggheads and Pie Trucks

In 1958, America's scientists and intellectuals could get away with projects like Argus in part because the space race had given them power and popularity. In Washington, physicists, engineers, and astronomers filled meeting rooms at the White House, the Pentagon, and on Capitol Hill and were sought for advice on every sort of decision. Lyndon Johnson was so eager to appear well informed and technologically savvy that, in the words of columnist Drew Pearson, one of the senator's main preoccupations became "wooing the eggheads."

The post-*Sputnik* period renewed the status that America's finest minds had enjoyed at the end of World War II when the secrets of the Manhattan Project were revealed and big thinkers became heroes. Of course a certain traditional wariness about intellectuals persisted. And for the general public, the IQ vogue was tempered in the middle of 1958 by the rumblings of a quiz show scandal, as even the handsome Charles Van Doren, described by TV columnist Jack Gould as the "bona fide egghead with enough sex appeal," fell under suspicion. But more than a year would pass before Van Doren would confess that he had cheated, and in the fall of 1958 many people continued to watch quiz shows and still looked to the intellectual class with admiration. At colleges and universities across the country, enrollment in science programs jumped by more than a third, with the biggest increases coming in the fields once deemed least enticing of all—math and physics. Even in high schools, the brains were suddenly getting the kind of respect formerly reserved for the jocks. High school officials in New Jersey, Tennessee, and even Texas announced that top-scoring students would be awarded letters for their sweaters, just like touchdown-scoring football stars.

If in a highly competitive society like postwar America the eggheads suddenly seemed worthy of admiration, it was at least partly due to the fact that they were engaged in a genuinely ferocious contest with the Soviets. The space race was somewhat like the Olympics, with national pride on the line in addition to the balance of military might. The power of rockets and nuclear weapons added what James Van Allen would call "a heavy dose of masculinity" to the otherwise nerdy image of the scientist, making the slide rule set somewhat sexy. Best of all, they were starting to look more like winners, and America loved winners.

Those keeping score when the one-year anniversary of *Sputnik I* arrived on October 4, 1958, noted that the United States had a four-to-three advantage in the number of satellites launched into orbit. Better yet, these birds flew much higher and stayed up much longer than the Soviet flock. *Sputnik*s I and II had already tumbled to Earth, leaving number three alone in the sky. Two of America's three *Explorer*s remained in orbit and little Vanguard was expected to stay aloft for centuries.[1]

These achievements were enough to reassure much of the American public about their government's effort to keep up with the USSR. Nevertheless, certain sources, including Henry Luce's *Life* magazine, continued to issue alarms about Soviet space advances, and some Democrats in the Senate found it hard to give up bashing Ike. In August a handful of Democrats had taken note of a long-term, government-sponsored study of the history of surrender in warfare and used the study to suggest the president was preparing for defeat. As they rallied the Senate to pass a better-dead-than-Red resolution, they ignored the fact that fellow Democrat Harry Truman had authorized the study and that it was an academic exercise, not a policy paper.

For his part, the current president had his press secretary James Hagerty declare the surrender issue "ridiculous" and then cited America's successes in space to support his party's standing with the electorate. "I think we have constructed a very steep curve of accomplishment," Ike told the press a few days before the *Sputnik* anniversary. "And I believe we have the biggest, strongest, finest body of scientists amply armed with money to do the job and that's that."

To back this statement, the president could point to new education programs he was working out with Congress for implementation in the coming year. The cunningly named National Defense Education Act would pump $250 million per year over four years into schools, colleges, and universities through a wide variety of programs. Elementary school teachers would get training and new materials for uniform science lessons such as "Mystery Powders" and "Batteries and Bulbs," which would become classroom staples. More significantly, the act created big new loan programs for college and university students. In time it would be regarded as one of the greatest education initiatives in history, improving schools nationwide, helping to establish premier scientific research centers, and opening higher education to millions of students who otherwise couldn't have afforded it.

The National Defense Education Act was approved by Congress days before the election, and the president and his GOP allies took pains to claim it as their own. With the nation struggling to emerge from the worst recession in almost twenty years, Ike's party needed good news. But the president, often an awkward partisan, didn't put much spin on the facts, whether he was talking about the NDEA or hailing recent American advances in space. It fell to his eager vice president to maximize the political gain. In a speech at San Francisco's Cow Palace arena, Vice President Nixon said U.S. satellite successes had made the space issue "deader than poor little Laika."[2]

Nixon was wrong. Space, and especially the issue of intercontinental missiles, would remain a vital political concern for years to come. And because there was more than one way to judge the race, no American should have claimed a definitive victory. Indeed, after the first year, the Soviets still had an enormous lead—roughly 4,200 pounds compared with about 100 pounds—when it came to the weight of the payloads put into orbit. In the mine-is-bigger spirit that characterized much of their Cold War propaganda, Soviet officials played up the disparity as they celebrated the anniversary of *Sputnik I*'s flight. One of their space scientists even described the American satellites that made it into orbit as no more significant than "unripe lemons."

Beyond size, the USSR could also claim superiority in the field of space travel. One dog successfully orbiting Earth, if only for a few revolutions, surely beat Wickie Mouse and all the other American rodents lost to the sea after brief rocket rides. And in late August the Soviets proved that they could also recover space travelers when two pups—Belyanka ("Whitey") and Pestraya ("Many Colored")—were shot to an altitude of more than 208 miles and parachuted safely to Earth in their sturdy sealed carrier. The dogs set an altitude record for animal flight, but more importantly, their survival and recovery put the Soviets firmly ahead in the competition to put a man in space.

In both countries, generals and politicians were eager to get men into orbit. The Soviets, who had continued to keep hidden the identity of the Chief Designer and conducted their space program in total secrecy, hinted that manned flights were their top priority. At the same time, in the United States, talk of manned missions could be heard in Congress, at the Pentagon, and at virtually every research center that had anything to do with space.

Prestige was a big motivation on both sides. Space, including the Moon and planets beyond, represented an almost unimaginable new frontier, and the nation that conquered it first would be regarded with awe. With this purpose in mind, President Eisenhower included human spaceflight, along with more satellites and eventually visits to the Moon, in a policy announcement based on the recommendations of his science advisory committee. Sounding a call that would eventually be echoed by the TV series *Star Trek*, this document cited humanity's age-old drive "to go where no man has gone before" as one of the main justifications for America's space ambitions.[3]

The second, and more urgent, justification offered for a big space program, including manned flight, was national defense. Military officials envisioned using rockets as vehicles to transport troops, to launch spy satellites, and to establish extraterrestrial bases. Plans were being developed in each of the services—Navy, Army, and Air Force—by officers and bureaucrats eager to claim certain space assignments and the dollars to fulfill them.

In just the Army alone, General Medaris's missile agency worked on a wide variety of space projects for various elements of the service. The Corps of Engineers naturally laid claim to the job of building manned military bases in space, whether the mission called for assembling orbiting stations like the one von Braun had imagined for Walt Disney or using bulldozers to build outposts on the Moon. In an article published a year after *Sputnik I,* a supporter called for construction of "military lunar stations" that would establish the U.S. Army's claim to function as a ground force — if that's the right term — throughout the universe.

Units based on the Moon and deploying nuclear-tipped missiles would enjoy two great advantages. First, since missiles wouldn't have to fly over the horizon, the weapons could be guided more precisely to a target. Second, the distance to Earth made them invulnerable to preemptive attack from the ground. In the time it took a Soviet attack to reach them, all of America's Moon-based weapons could be launched.

While Army engineers imagined barracks and mess halls and rocket launchers on the Moon, the Army Signal Corps won permission from the Pentagon to develop a missile that would gather television pictures to evaluate battlefields. The Army Transportation Corps got involved in designing a rocket to deliver supplies and troops via a swollen Redstone rocket that would release a large capsule designed to float down to Earth with the aid of a huge parachute. According to the design, once it landed safely, the men inside could scamper out, or the rocket could be used to supply ammunition, rations, or other cargo to troops waiting nearby.

If the Army seemed to be working overtime to imagine its role in space, it was partly because the Air Force, an upstart in the defense community, was gaining in the competition to build powerful missiles. Despite all the efforts to quell this contest for the sake of the federal budget, it had continued to blaze, with the results announced in the press with the urgency of World Series game stories. In May 1958, *The New York Times* reported that the Eisenhower administration cut production of the Army's mid-range Jupiter missile while it ordered additional numbers of the comparable Thor from the Air Force. A few months later came the news that the Air Force's Atlas and Titan missiles, which were designed to match the Soviet's long-range R-7, were nearly ready for production.[4]

The encouraging forecast for Titan and Atlas was announced in mid-September by Simon Ramo, chief missile designer for the Air Force, who appeared on a nationally broadcast TV and radio program called *College News Conference*. Ramo was a brilliant and widely respected scientist who rarely spoke in public forums. For these reasons, when he noted that, thanks to the Air Force missiles, "we now may be ahead of the Russians" the words carried weight. His company, Ramo-Woolridge, was a main scientific resource for Air Force ballistic missile programs. What Ramo couldn't say, and perhaps didn't know, was that the balance of power was shifting to America's favor in part because Soviet claims were withering under close scrutiny. Defense and intelligence experts were finding no evidence that the USSR had a combat-ready ICBM. In fact many doubted the Soviets could deploy one anytime soon.

Simon Ramo's evaluation, made without reference to Soviet shortcomings, reflected the increasing confidence within the Air Force missile group, which more and more seemed to represent the future of America in space. While the Army was still working with machines descended from the German V-2, much of the technology in the Atlas and Titan was new and represented greater efficiency and elegance of design. But even more importantly, because the main business of the Air Force was flight, it had a great advantage when it came to the more glamorous task of putting a human being in orbit.

Throughout the 1950s, the Air Force had conducted extensive work on the human aspects of space travel at places like the Aeromedical Research Laboratory at Holloman Air Force Base in New Mexico and the School of Aviation Medicine at Randolph Field, Texas, where Airman Donald Farrell had spent seven days cooped up in a metal box for his simulated mission in space. The other service with a major interest in flight, the Navy, had its own aviation medicine specialists and conducted extensive tests, including centrifuge rides that subjected both men and animals to forces up to forty times gravity.

But while the Navy did important basic research on flight, most of the famous daredevils like John Stapp and David Simons, who had strapped themselves into rocket sleds and ridden ballons to the edge of space, were Air Force men. With their experience, and a corps of test pilots to tap, the Air Force seized the lead in the quest for manned orbit with a post-*Sputnik* proposal called MISS, for Man in Space Soonest.

By the summer of 1958 the Air Force and its allies in other agencies had begun to work on a scheme that would use combinations of existing rocket motors to orbit first monkeys and then men inside metal containers—initially called "space cabins"—shaped a little like light bulbs. The vehicles represented the simplest option for quick space travel. Built to be as light and foolproof as possible, they would have no wings, fins, or rudders. They would be little more than protective cages equipped with heat shields capable of withstanding a launch and reentry to splashdown in the ocean. Creature comforts inside the cabin would be minimal, with life support systems—heating, cooling, air supply, and so forth— capable of sustaining a passenger for no more than two days.

The space cabin was, in the eyes of pilots, the ugliest of the alternatives. They preferred the idea of actually flying a plane into space. For years they had been given reason to hope for a rocket plane trip to space, as the Air Force and the National Advisory Committee for Aeronautics had been testing prototypes that reached ever higher altitudes and speeds. Launched from beneath a high-flying bomber aircraft, the Bell X-2 Starbuster piloted by Iven Kincheloe had set an altitude record of 126,200 feet in September 1956. However, three weeks later pilot Milburn Apt died in a crash after setting a world speed record with the same swept-wing aircraft. As of late 1958, no rocket plane was capable of space orbit, and developers were still several years away from building one that might serve this purpose better than a capsule atop a rocket. For the short term, at least, the American astronaut would function as a man-in-a-can, not a flying ace.

The dumbed-down duty that would fall to the first astronaut made the job less appealing to macho test pilots. Even so, in the summer of 1958 many of these gung ho fliers, including Neil Armstrong and Scott Crossfield, would be assigned to the first official list of astronaut candidates. (Kincheloe would likely have joined them, had he not been killed in July when the engine on a jet he was flying failed during takeoff.) These men had proven their physical and psychological strength, the two main qualities required to handle the kind of stress they would endure rocketing into orbit and coming back to Earth as passive test subjects. They were selected with the expert help of Stapp, Simons, and other veterans of space medicine experiments. Having experienced the kind of stress the first astronauts would endure, Stapp and Simons recommended ways to reduce the strain on space cabin passengers with multistage rockets, and they pressed for a long series of animal tests to be conducted before any man would be asked to risk his life atop a rocket.

Considering the uneven record of launches at the Cape—explosions on the pad, debris in the Banana River—the danger of a rocket ride was mortally obvious. But whatever the outcome, the first astronaut would be a hero in an instant, and a historic figure for eternity. Similarly, the branch of the armed services that got into orbit first would have a special claim to future manned missions in space. No doubt this was on the mind of General Curtis LeMay, by this time the Air Force vice chief of staff, who was one of the main advocates for MISS in Washington.

Though Air Force officials at the Cape rarely saw LeMay and considered him an airplane man and not a rocketeer, they understood that the general was determined to claim everything that flew for his branch of the military. "I never saw him once, and he wasn't much for missiles but he wore the blue [Air Force] suit," recalled Ken Grine, then the press officer at the Cape Canaveral missile range. "The doctrine was that it if went into the air, it was ours."

Remarkably, LeMay had initiated a claim on space for the Air Force twelve years earlier when, in the aftermath of World War II, he commissioned a study called *Preliminary Design of an Experimental World Circling Spaceship.* The first report ever produced by the RAND Corporation think tank, this document showed how a satellite could be put into orbit and how it might be used for surveillance, communication, and even missile guidance. It also argued that the Air Force should lead America into space as an extension of its airpower mission.

The RAND report put space on the Air Force agenda and it remained there as many of its authors' predictions, including the political gains made by the Soviets when they got there first, came to pass. Then in 1958, when the Defense Department called for manned spaceflight proposals, LeMay made sure that his service lobbied hard for the job. The Air Force also maneuvered to join the satellite business in a dramatic fashion, targeting the Moon for fly-by

·missions, which had long been considered the first step toward landing and establishing a lunar base. In all, the service proposed a variety of projects to be pursued in the coming five years with a cost
estimated at $1.7 billion.

Though conducted quietly, the Air Force push signaled a change in
the interservice competition. "Medaris and the Army had put up the
idea of going to the Moon but the Air Force was working more closely
with the people putting together the civilian space program," recalled
Grine. "You could feel the change, like one group had the advantage
and was starting to move."

The first Air Force Moon shot would be attempted in August with
Thor-Able, carrying a satellite called *Pioneer 0*, which was rigged to send
infrared photos and radiation readings back to Earth. The event would
long be remembered because it marked the first nearly live national
television broadcast of a launch from the Cape. Until this point, video
cameras had been barred from the Cape and all TV broadcasts showing
rockets in flight had depended on old-fashioned film. After a launch,
reels would be packed and then flown out for processing, usually in
New York. In the time this took, the military could gather information
and put together an explanation for what went right or, just as often,
what went wrong.

In general the Air Force tried to be more accommodating to the
press, in order to enhance its public image. (Army officials called the Air
Force public relations policy "wide open.") The blue suits treated reporters to free trips to various bases and even rides in exotic airplanes.
These outings made an impression. Wickie Stivers, for example, would
treasure forever photos taken when she was suited up for a jet fighter
ride. Prior to the Moon shot in the summer of 1958, the Air Force had
loosened controls on reporters, announcing that the networks could use
video cameras and then go on the air more promptly with images of the
blastoff and early flight.[5]

When a letter noting the new rules arrived at the local NBC bureau,
young Jay Barbree took a lawyer's eye to it and noticed that no specific
time delay was required before images of the launch could be rebroad-
cast. Recalling how radio call-in programs used two tape recorders to
create a delay that would allow the bleeping of profanity, he suggested

that technicians use two videotape machines to broadcast events with a similar delay just seconds after they occurred. Higher-ups at the network liked the idea.

The launch was scheduled for the morning of August 18. On the night before, a mobile broadcast studio nicknamed the Pie Truck, because it was a converted bakery delivery van, made its way from the network's Jacksonville affiliate to the Cape. Painted a soft cream color and decorated with the station's call letters—WFGA—the truck was one of the first rolling studios equipped for live broadcasts. The NBC crew chose to bring it south under cover of darkness so that competitors wouldn't see it and catch on to what they planned.

After a bit of confusion at a security gate the van and the men inside were admitted to the test range. They were ordered to park behind the building designated for the press. After the men got out of the truck, they opened the big back doors and began unloading equipment. Studio cameras weighing up to 200 pounds were hauled up ladders to the rooftop of the press building. A big microwave dish was then taken out of the truck, bolted to its roof, and pointed to a receiver on a tower nearby.

The next morning there was nothing that other TV networks could do to match NBC's technology. The rocket was fired at 8:18 A.M. into a bright blue sky. On the rooftop where the press was posted, the NBC men braced themselves to tilt back the heavy camera so it could follow the racing missile into the northeast sky. They looked like the famous statue of the flag raising at Iwo Jima. All seemed fine as the missile climbed toward the Moon, but seventy-seven seconds into its flight the Thor exploded. The pieces fell into the ocean from an altitude of ten miles.

With the tape-delayed video signal bouncing by microwave all the way to his network, NBC reporter Herbert Kaplow was on the air nearly live with the story. Viewers got to see everything, from the first puff of vapor at ignition to the midair explosion just seconds after they occurred. Air Force officials didn't discover that the nation had witnessed their failure until they met with the cameramen from the other networks. "They were arranging to fly the film out when one of the guys said, 'What the hell for? It's already been on the air,'" recalled Herbert Gold, who had been a producer for the Pie Truck crew.

Although it irritated government officials, the broadcast, which cost NBC almost $30,000, began a tradition of live coverage that would continue at the Cape for decades. Rocket launches, with all their delays and drama, became an entertaining staple of the American news diet. In fact, both NBC and CBS were on hand with video cameras when the Air Force fired a second shot at the Moon at 4:42 A.M. on October 11.

This time all the engines worked and *Pioneer I*, which was shaped like a toy top, quickly raced farther into space than any man-made object had ever flown. Although the Thor was the most complex rocket ever launched by the United States, and thus required the most intricate mechanical and electrical choreography, it performed almost flawlessly. This was true, even from a public relations standpoint, as the rising rocket carried the letters U-S-A-F for every American voter and taxpayer to see.

A single shortcoming, however, denied the Air Force its ultimate success of a lunar fly-by. This Thor-Able was just a shade too slow and failed to reach the velocity—25,000 miles per hour—required to break Earth's gravitational grip. After traveling up 79,000 miles, it slowed, stopped, and began to fall down. Less than two days after it had departed Earth, *Pioneer I* burned up in the atmosphere over the Pacific Ocean about 2,000 miles west of the point where Chile and Peru share a border.

In a display of the Air Force's PR acumen, *Pioneer I* was widely promoted as a big success even though it failed to reach its goal. Officials noted that the satellite had set impressive speed and altitude records. It also had provided new data on conditions in deeper regions of space, which showed safer radiation levels once the craft exceeded a point 20,000 miles from Earth. This news was encouraging for those who contemplated manned flights and hoped to spare astronauts excessive exposure. *Pioneer*'s data also gave James Van Allen and other physicists evidence that the Earth's magnetic field dipped down at each of the poles, attracting radiation in the same pattern that a bar magnet attracts iron filings.[6]

Pioneer I's achievements moved *The New York Times* editorial page to pronounce the mission "A Glorious Failure" that was as important as the first *Sputnik* and to note "how far American rocketry has advanced since that first unhappy Vanguard failure only last December." The success also reminded the world that the USSR had announced no advances in rocketry or satellites for months. Except for vague hints made at international conferences and in journals, Soviet scientists and officials offered no information on their plans and the isolation of their rocket facilities meant that failed attempts could go unnoticed.

In truth, with the support of their government, the Chief Designer and others in the USSR had been working hard on nuclear missiles, manned spaceflight, and the challenge of sending satellites to the Moon. Even Khrushchev had gotten into the process, proposing that rockets be placed in protective buried "silos." At first the idea was rejected as unworkable, but when a similar proposal was made in the United States the Soviets went back to the chairman's idea and began to develop it.

While the silo idea would lead to important practical results, much of the space work the Soviets considered in late 1958 was quite fanciful. The Chief Designer was intrigued, for example, by a scheme for bringing a space capsule back to Earth by equipping it with a helicopter rotor and blades. Another capsule design called for an umbrella-shaped air brake. But for the most part the Soviets conducted serious efforts to improve their rocket power, communications, and guidance so they could match or beat their American competitors. Failure, or rather, admitting failure, was not an option. Between September 24 and December 3, they tried three times to crash a silvery, ball-shaped scientific satellite into the Moon. Each of the modified R-7 rockets was launched from Baykonur. The first two exploded shortly after liftoff. The third shut down less than five minutes into its flight. According to available records, these failures were not detected by American intelligence and would not be acknowledged for years.[7]

★

In at least one area of space science and engineering, America chose openness over secrecy. Though the fact was little noticed, the *Pioneer* satellite that flew a third of the way to the Moon was the first ever launched with the guidance of the new National Aeronautics and Space Administration. A civilian agency that eventually would assume the nation's most visible space projects, NASA had been created in a rush.

Proposed by the president on April 2, the legislation creating the agency was subjected to hearings, voted on by both houses of Congress, and signed into law in fewer than 120 days.

NASA's nucleus would be the offices and laboratories of the National Advisory Committee for Aeronautics. Born in the Wright brothers' era—the Kitty Hawk flight was depicted on its official seal—NACA had played a big role in designing the aerodynamic elements of propeller and jet aircraft. With partners in government, science, and industry, NACA had also produced important technical advances in rocketry. Although NACA had 8,000 employees and research laboratories in Ohio, Virginia, and California, these resources were hardly enough to sustain a national space effort. For this reason, NASA was expected to acquire or develop more labs and engineering centers until it was a much larger and more capable bureaucracy.

Responsibility for cobbling together the big new agency fell to its first administrator, T. Keith Glennan, a North Dakota–born engineer with a remarkably varied background for a preeminent egghead. After earning a degree in electrical engineering at Yale, Glennan found work in the film industry as new technology made talkies the standard. He wound up managing movie production for the Paramount and Samuel Goldwyn Studios, before World War II brought him into weapons research. Afterward he returned to the job of president of the Case Institute of Technology, which would soon merge with Western Reserve University. The new Case Western University rose to national prominence as a center of scientific and engineering research.

Though he was a board member of the National Science Foundation, on the day he was offered the administrator's job Glennan knew no more about the space race than any avid newspaper reader and little of what NASA was supposed to be. He took a few days to decide, but eventually agreed to leave his quiet academic post for one of the most visible assignments in all of government. Temporarily housed in the historic Dolley Madison House, in sight of the White House, NASA headquarters was immediately at the center of national politics. Glennan was pressured from all sides to appoint hack candidates to certain jobs— some were refused, others accepted—and he was forced to learn, on the fly, the true extent and limits of his power.

Under the legislation that created NASA, Glennan had until the end of the year to find bits and pieces of other government agencies to add to the NACA base to create a real civilian space program. As long as the president approved, he could select almost any asset he wanted and annex it without the assent of Congress. Come January 1959 he and the president would lose this authority, and any attempt to grab a federal office, bureau, agency, or lab for NASA could be challenged in the House or Senate.

Early in his shopping, Glennan went to Huntsville, where he found much to admire in Wernher von Braun's research and engineering team and a great deal to dislike in John B. Medaris. "He was addicted to spit and polish," Glennan would recall, "and determined to beat the Air Force" for the lead role in military missilery.

In the short time he had studied the issue, Glennan was convinced that Medaris "didn't have the cards" to win the contest with the Air Force and his best engineers and scientists—both German and native-born—would enjoy a much brighter future of service to their country as part of NASA. He also became certain, based on talks with von Braun, that he and his group would happily change agencies if it meant they would get ample funding for space exploration and be free from military work. "All we really want," von Braun had told Glennan, "is a very rich and generous uncle."

Year later, Glennan would describe his visit to Huntsville in a sketchy way, recalling that Medaris treated him in a "somewhat cavalier fashion." Medaris and his top officers had reason to feel equally mistreated by the man from NASA. One would recall that Glennan repeatedly asked why the Army, a ground-based force, had any involvement at all with missiles and satellites. Answering his own question, Glennan decided that NASA should acquire both the von Braun rocket team and the Jet Propulsion Laboratory in Pasadena, which had done so much Army satellite work.

The transfer of JPL and half the Army missile group (including von Braun) to NASA seemed reasonable to Glennan and to President Eisenhower and his science adviser, James Killian, when the three men discussed it at the White House. After all, you couldn't have a space program without proper rockets and vehicles to explore the heavens. Besides, the Air Force was clearly winning the missile competition within the Department of Defense. Even though Ike warned of possible Army resistance, Glennan confidently went to the Pentagon in mid-October to discuss the transfer with assistant secretary of defense Donald Quarles. The first sign of trouble came when Quarles suggested that Glennan work things out with the secretary of the army, Wilbur Brucker.

Glennan walked to the secretary's office, where he found Brucker sitting with two generals, Arthur Trudeau and John Hinrichs, "who sat there like wooden horses throughout the entire conversation, never uttering a word," recalled Glennan. As the NASA chief began to talk, in a halting way, his spirits sank. An irate Brucker, whom Glennan later called "one of the most stupid persons I have ever met," would have none of it. Von Braun and the ABMA were simply not going to be separated from the Army, he decreed, and there was no point in discussing it further.

On the day when Glennan visited the Pentagon, Bruce Medaris, who happened to be in Chicago, got word of Glennan's proposal for taking over von Braun's team. He assumed the Air Force was behind the idea and feared that if Glennan got time to press his case, the president—the "Great White Father" to Medaris—would make it happen. Far more experienced than Glennan when it came to military and political fights, Medaris decided to respond with the swiftest weapon at hand—a leak to a friendly reporter.

Mark Watson of *The Baltimore Sun* was, in Medaris's view, "honest, reliable, objective, patriotic and thoroughly dependable." So dependable, it turned out, that he met Medaris soon after the general got off a plane from Chicago. The next day's *Sun* rang a front-page alarm about the potential destruction of a great national asset by a new agency—NASA—that was prepared only to use a fraction of its talent. Other press outlets picked up the story and soon the entire country was reading about Glennan's grab and how it threatened the Army's most important asset. An editorial writer for *The St. Louis Globe-Democrat* published a typical piece. Declaring "weapons come first," it described the transfer of the Army group a potential "calamity." It concluded, "If NASA can't take a close-up of the man in the moon without stripping the scientific cupboard bare it would be better if NASA closed up shop."

With a handful of generals joining him, Medaris managed to use the press to fight Glennan's move and put pressure on the White House to change direction. Given Eisenhower's status as the Army's hero general of World War II and armed forces commander in chief, a challenge of this sort could be viewed as unseemly. However, it was really just the latest of many conflicts between the soldier who saw himself as a citizen first and a vast military bureaucracy that he sometimes struggled to control.[8]

The most open drama of this type, begun when General James "Slim Jim" Gavin had resigned in a huff in January, had continued throughout the year. In August Gavin had published a book, *War and Peace in the Space Age*, which quickly joined *Lolita* and *Kids Say the Darndest Things!* on *The New York Times* best-seller list. In this book Gavin continued his attack on the Eisenhower military scheme and painted a radical view of the nation's defense needs. Declaring that the Soviet Union was already at war with the West, he described a not-too-distant future when the entire planet was a potential field of battle for an "Earth War" fought by weapons and troops—including Sky Cavalry—deployed via outer space.

The president had answered Gavin's charges directly, saying, "I am quite certain that the Defense Department's programs are not only quite adequate but really are generous." But while Eisenhower could deal with Gavin with a few words to the White House press corps, the larger uprising over the Huntsville missile group was a more serious problem. As the generals gathered public and political support, the administration decided to back away from Glennan's proposal, at least in the short run. The Huntsville group was left intact and would work as a contractor for NASA. Glennan, who had been shocked by the Army's response to his plan, wouldn't raise the issue again until 1959, after his ally Herbert York was named director of research and engineering for the entire Department of Defense.

In the meantime, Glennan acquired the Jet Propulsion Laboratory for NASA and moved quickly to approve a man-in-space program called Project Mercury. He also ordered work to begin on a huge rocket with 1.5 million pounds of thrust to serve as the launcher for space projects in the mid-1960s and beyond. The 1.5 million figure was chosen as much for its psychological effect as anything else, Herbert York would later explain. No one knew exactly what kind of rocket would be required, noted York, but big round numbers had their appeal.

The start of Mercury and the big rocket program, which would lead to the development of the famous Saturn booster, marked America's clear transition from somewhat scattered, mostly military-based space exploration efforts to a single, highly organized civilian program under NASA. Soon, under a different president, this fast-growing agency would be asked to race the Soviet Union to put men on the Moon. NASA's size and central management would prove essential in this competition, and would affirm the wisdom of the space policy set by Congress and President Eisenhower in *Sputnik*'s wake. The shift would ultimately allow the United States to claim dominance in space with ever more impressive achievements. But old hands at the Cape, where ducks were hunted with compressed air cannons, would miss the days of adventure when rocketry was practically their private domain.[9]

ELEVEN

The Monkey and the President

When President Eisenhower learned that two Air Force generals had been spouting off to the press about their plans to conquer space, he couldn't understand why these men talked so much about something that had a 50-50 chance of exploding in their faces. He pounded on the table in the Cabinet Room and demanded to know, "Why don't *they* make those two generals shut up?"

The scene would forever amuse presidential adviser Herbert York, who was at the meeting. The commander in chief was asking the highest-ranking officials in the Air Force, also at the meeting, why some mysterious "they" were so powerless. "They" were all sitting at the same table, York would recall, and as he saw it, the blabbermouths got away with their loose talk because "the Congress, the press, and ultimately the people" wanted to hear what they had to say.[1]

In fact, most of the military men who led the space race for the American side had built constituencies in the Congress, the press, and industry that provided them real protection and power. General Medaris had proved this point when he had beaten back T. Keith Glennan's effort to raid Huntsville for von Braun and his missileers. This was the real reason for the president's frustration. Some things were beyond his control. It also explained why the president chose to keep secret the big surprise planned for the end of America's first year in space.

The idea was proposed by Roy Johnson, a political appointee who was head administrator of the Pentagon's Advanced Research Projects Agency, where York served as the chief scientist. Johnson, who had taken office April 1, had been an odd choice to run a big new science and technology agency. He came from General Electric, where he had been the vice president in charge of the appliance division, which made and sold everything from electric clocks to washing machines. When he was appointed he confessed to having no background in science or military matters, but he reminded the press that he was an organization man, the type of leader who could put together a complex bureaucracy and make it work.

As a modern executive, Roy Johnson, who had set records for appliance sales at GE, clearly understood the value of positive images and the importance of public perception. Public relations pioneer Edward Bernays, who had worked for GE, had recently codified the practice of manipulating opinion in a book called The Engineering of Consent. Bernays's ideas, which were based in part of the work of his uncle, Sigmund Freud, had become part of the business canon. Indeed, by 1958 public relations was a large and respected discipline engaged in almost every corner of society. President Eisenhower had almost doubled the number of public affairs specialists in the bureaucracy, and "news management" had become an established political tool.

Public relations must have been on Johnson's mind in May when he visited the Convair Corporation's Atlas plant near San Diego. The Soviets had just launched Sputnik III—the largest satellite in space— and like many high government officials Johnson was irritated. "We've got to get something big up," he told his hosts.

In the Convair boardroom, Johnson learned that under the right conditions the Atlas missile would be capable of going into orbit. He then heard a proposal for a shortcut to the heavyweight satellite title— putting an entire Atlas rocket, more than four tons of hardware, into space. The idea was so exciting that before he left the Convair complex, Johnson went to inspect a missile that was available for the job, an Atlas 10-B. With a flourish, Johnson grabbed a pen and wrote his name on it.

Of course it wouldn't be enough to throw dumb metal in the air. The missile-cum-satellite would have to serve a technical purpose. Together the contractor and its Air Force partners proposed to load radio equipment into the body of the missile to create the first space-based communications relay station. It would accept voice and data messages from Earth and then broadcast them back. Technically, it wasn't much of an advance over previous satellites, but that wasn't the point. The voice messages could be transmitted so that listeners around the world would hear them. People wouldn't be able to resist this novelty—a voice from the heavens—and everyone who tuned in would be reminded that it was the Americans who were biggest and best in space. It didn't hurt, from a PR standpoint, that the endeavor could be called Project SCORE, for Signal Communication by Orbiting Relay Equipment.

Roy Johnson embraced SCORE immediately, but York, his chief scientist, wasn't impressed. "It was pretty phony," recalled York, decades later. "The idea was for the Air Force to put the whole carcass of an Atlas up there with communications devices to give it some purpose. Johnson didn't know technology, didn't know that this was not that impressive, but he thought he knew public relations."

A large man with a strong voice that brimmed with self-confidence, physicist York was rarely shy about speaking his mind. He was accustomed to the candor of scientific debate, and at most meetings could be relied on to take his pipe out of his mouth and dispense with a silly idea with just a few words. But in this case, since he was just getting to know Johnson, York decided to hold his criticism. "I didn't want to carp," he recalled, "so I went along."

At the White House meeting, the president said he liked the SCORE concept but was determined to avoid a public humiliation like the first Vanguard attempt, should the thing fail. Johnson and York were ordered to pursue the plan as a top secret project with the smallest possible number of people. A handful of experts at Convair would be in on the plan, along with a few Air Force officials and a limited number of technicians, including those who would build the radios at the Army Signal Corps center at Fort Monmouth in New Jersey. The money for SCORE would come from ARPA's $340 million start-up budget. Final authority for launch would rest with ARPA and the White House. Blabbermouth generals would be kept in the dark.[2]

★

SCORE, led secretly from the White House and ARPA's office in the Pentagon, represented a further shift in the nature of America's space adventures. Although General Medaris would probably have disputed the point, the Army was quickly being outpaced. Wernher von Braun seemed to signal that he understood this in the fall of 1958 when he met with York to discuss his future. Weeks before, he had issued a public statement opposing the "dissolution" of the Army space group. But with York he seemed far more open to change. "He wanted to go where the money was," recalled York. "He was interested in continuing his work, being of use, and getting the most financial support possible."

It seemed to York, and probably to von Braun too, that sooner or later most of the ABMA missile group would be folded into NASA. In the meantime, von Braun and Medaris would focus on a Moon shot that had been approved for early December. They needed a success to restore their reputation, which had been weakened with summer's failed *Explorer V* satellite attempt and the spectacular explosion of a Jupiter that occurred in early October. That particular missile had gone out of control and was destroyed by the range safety officers when it was just a few hundred feet in the air. Debris had rained on a wide swath of the Cape, threatening but not hitting the pad where the Air Force was preparing to launch its next Atlas.

The plan for the December Army Moon mission called for a modified Jupiter missile called Juno II to throw a thirteen-pound satellite deep into space, where it would then come under the gravitational influence of both the Moon and the Sun. According to calculations made by the Army group, the combined gravity of these two bodies would tug the satellite past the Moon and then into a solar orbit. The science aspect of the mission would be done by radiation detectors that would collect data to be radioed back to Earth.

The rocket blasted off just after 12:45 A.M. on December 6, and climbed with enough smoke, flame, and thunder to suggest it had the power to achieve its goal. All four stages fired properly, and as the satellite rose higher, it accelerated to a speed greater than 23,000 miles per hour. At that rate the satellite, called *Pioneer III*, would reach the Moon in half a day. (One month earlier, a satellite called *Pioneer II* had failed to reach orbit.)

But the flush of success that came when the rocket safely departed the Cape and seemed to perform as programmed, gave way to doubt as tracking crews checked the satellite's flight after the last rocket stage ceased to burn. *Pioneer III* was traveling at an extraordinary rate of speed, but it was not going quite fast enough. One rocket motor had cut off too soon. The spacecraft's velocity was low by about 900 miles per hour. By 3 A.M., the launch crew knew their ship was not going to make it.

Like a shell fired from a mortar, *Pioneer III* was going to fly high and then descend, very rapidly, back to the ground below. The Army bird wasn't even going to reach the same altitude—more than 70,000 miles—that the Air Force Moon shot had attained in October. Still, the satellite would prove useful. As it climbed, it sent out data on the layers of radioactivity in the near regions of space. And more information would be gathered on its return trip, before it was destroyed by atmospheric friction.

William Pickering of JPL emphasized the positive scientific value of the failed Moon shot when he briefed the bleary-eyed reporters who gathered to hear him in Washington, D.C., sometime around 4 A.M. When added to existing information being analyzed by James Van Allen and others, the information from *Pioneer III*'s superior instruments would confirm that two belts of radioactivity encircled the Earth, not one. Eventually, the information from the satellite would also help determine ways to manage the radiation hazards of human space travel.[3]

Pioneer III's data mattered because the real future, at least where public acclaim and interest were concerned, lay in putting human beings in space. For more than a century space adventure had been a popular fantasy, and the excitement of the satellite competition had only increased Americans' interest in the possibilities of travel beyond gravity's grasp. Suddenly the impossible seemed likely and brought new speculation in the press almost daily. In California the most avid consumers of space information could even enroll in a so-called Space School offered by the extension division of the state university in San

Francisco, Los Angeles, and San Diego. The visiting lecturers included, among others, von Braun and Pickering.

Inevitably, as adults talked about sending humans on long space voyages, the question of sex arose, and at least one real expert suggested that romance might be a ready cure for an ailment that could strike most astronauts: loneliness. As a psychologist on staff at Convair noted, "There is ample evidence that man is better able to survive the boredom and isolation of space travel by having his wife along with him." A baby who grew to term and was then born in space, added Donald Conover, Ph.D., might be specially suited for life in the weightless cosmos.

The prospect of manned—and wo-manned—space missions excited the imagination in Hollywood too. In late 1958, those who couldn't wait for the real thing could gape at a campy version for the price of admission to *Queen of Outer Space*, which featured Zsa Zsa Gabor as one of the miniskirted inhabitants of Venus. (She plays a scientist named Talleah.) On the day when *Pioneer III* was launched, Zsa Zsa's *Queen* shared the bill at a Cocoa Beach theater with Doris Day and Richard Widmark in *The Tunnel of Love*. For verisimilitude, Edward Bernds, the director of *Queen*, used film of actual rocket launches in his final cut. The blastoffs were real, although a careful viewer might notice that footage of a V-2 and an Atlas were mingled for a single scene. Bernds would follow this effort in 1962 with *The Three Stooges in Orbit*.

Given the free, televised spectacle of real and dangerous rockets roaring off the pads at Cape Canaveral with the imminent possibility of disaster, it's a wonder anyone felt inclined to pay for the fictional version, even if it did portray an Amazonian-style Venusian society. Great amounts of money and talent were being poured into actual missions that could prove to be highly entertaining. By December 1958 a dozen different companies had submitted to NASA designs for a capsule to carry a man on a rocket-powered flight. At the same time experts in physiology, medicine, and even psychology were settling on the ideal qualities of the first astronaut. Before year's end the government was scouting for men who fit stringent criteria, including a college degree or equivalent, test pilot training, excellent physical condition, and 1,500 hours of flying time.[4]

The test pilot requirement had more to do with the expected astronaut's physical and emotional makeup than any flying skill. The capsule, as envisioned in the final design, would be more of a projectile than a maneuverable craft, and except for employing jets to control pitch and yaw during part of the flight, the astronaut's main assignments would involve staying calm, making reports via radio, and getting out of the capsule after it splashed into the ocean.

Theoretically, the job of the first astronaut could be made so simple that a monkey could do it. In early December, a team of medical researchers led by Dr. Norman Lee Barr prepared a one-pound astronaut who would make this point as the first primate in space. This nine-month-old squirrel monkey named Gordo had been trained with half a dozen other natives of the South American forest. He won the honor of being first in flight because he learned to fall asleep soon after he was placed in the confines of a miniature space capsule. This habit earned him the nickname of Old Reliable, because his slumber meant he didn't disturb the sensors that monitored his breathing, heartbeat, temperature, and other bodily functions.

As one of Gordo's chief sponsors, Dr. Barr was a modern macho scientist cast in the mold of the great explorers. One of the Navy's first flight surgeons, he was so busy that the service kept a multi-engine, R4D transport plane ready for his use at all times. Though based at Bethesda Naval Hospital, Barr also worked at the Navy's School of Aviation Medicine and Research in Pensacola—where Gordo was trained—and lectured on the medical aspects of flight around the world. In the summer of 1958 he had flown to Egypt just to drop off his eighteen-year-old son, who would spend his school break catching kangaroo rats to be used in experiments back home.

"My father wanted to send them into space because they produced no urine and that eliminated certain problems," recalled Norman L. Barr Jr., many years later. "Sending me to Egypt was the kind of thing he just did. I remember that when he wanted to learn more about rockets, he just brought Wernher von Braun to our house."

Gordo's flight would mark a high point in Captain Barr's career. Barr flew to Patrick Air Force Base well before the countdown began. Liftoff was scheduled for about 4 A.M. Saturday, December 13. Hours before, a group led by Barr and another Navy doctor named Ashton Graybiel brought their subject to the launch complex and spent time there swinging him by the arm and bouncing him up and down. An hour or so before ignition he was wired with sensors, dressed in a little helmet, and slipped into a glass and metal canister a little bigger than a can of tennis balls. Inside the can his body was supported by a bed of molded silicone much like the chairs future human astronauts would occupy in their space capsules.

Once sensors were switched on, instruments at one of the ground facilities flashed readings of Gordo's heart rate and other functions, showing that all was well. The canister was then walked out to the waiting Redstone missile—its mission designated Bioflight 1—and then slipped inside the nose cone.

"The monkey was far less excited than we were," wrote Barr, shortly after the flight. "True to form, he fell asleep."

The deafening sound of the Jupiter's Rocketdyne engine woke Old Reliable, but his heart rate and breathing were surprisingly steady. Thirty seconds into the flight, as the speed increased, so did the g-force bearing down on Gordo's body. He held his breath, and then resumed breathing slowly, sighing as he exhaled. His heart pounded faster, but the rate of increase was only about 10 percent. For two minutes more, the rocket continued to burn and the g-force built to the point where Gordo's body felt ten times heavier than normal. Then, as a controller in the blockhouse announced "Burnout," the engine suddenly quit and the nose cone went into free flight, where the monkey in his can would be suddenly weightless. Barr wrote:

> As the seconds ticked away we became almost incredulous.
> Immediately after burnout, Reliable's respiration dramatically returned to normal. Within forty-five seconds his pulse rate leveled off. Watching the dials, one of the doctors kept repeating. "This is amazing, This is amazing."

For eight minutes the monkey floated in space, apparently relaxed and unaffected by the absence of gravity. Microphones even picked up his making happy, chattering sounds. Barr was thrilled by Gordo's reactions because, on a physical basis, he served as a more than reasonable stand-in for a human being. (Much better, he thought, than a dog.) In fact, almost the only differences between a monkey astronaut and a man were size, obviously, and intellect. Without offending Gordo, it was safe to assume a human rocket-rider would reflect and analyze the experience more fully and therefore show a little more stress. After all, he would know the launch was coming, understand it was perilous, and anticipate every moment, right down to landing in the ocean and hoping for rescue.

Once Old Reliable's capsule reached apogee—at roughly 300 miles high—and began to descend, deceleration took over with a vengeance. At top speed, the falling capsule was moving at 10,000 miles per hour and Gordo felt forty times his body weight pushing him down. Remarkably, his pulse rate rose only slightly, and while his breathing again became labored, he stayed conscious.

The main purpose of Bioflight 1 was to measure Gordo's response to the launch, weightlessness, and return to Earth. Even before the capsule landed, the flight was the biggest success yet for the American study of biology in space because it showed "with virtual certainty," as Barr reported, that the human body would handle well the demands of spaceflight. But this success was not enough. For reasons of sentiment as well as science—they were curious to study his organs—Gordo's handlers wanted him back. To this end, they had rigged the nose cone with two parachutes, which opened at an altitude of 8,000 feet to slow the falling capsule, as well as a balloon to keep it from sinking once it hit the water, and fluorescent dye to color the ocean bright yellow so that searchers might readily find the capsule.

The parachutes opened as expected, and the passenger's heart rate registered his relief as the g-force subsided. But then two problems interfered with his recovery. First, the capsule overshot the expected landing spot, where ships waited, by about five miles. Second, the nose broke and flooded with water. This was the conclusion reached when the Navy found nothing in the impact zone. Gordo, first American primate in space, gave his life for the cause.

Although Old Reliable never received the hero's welcome he deserved, *Life* magazine printed his image as part of a two-page spread.

But Gordo's moment would be brief. In a matter of days Project SCORE would seize the world's attention and the first primate in space, an astronaut's cousin, would be forgotten by all but the most avid space history buffs.[5]

★

For six months, scattered engineers, technicians, and bureaucrats had worked in secret to turn a scheduled routine test of an Atlas B missile into an actual spaceflight. The chain of men who knew what was happening included some top officials, like General Donald Yates, who ran operations at the Cape, but several hands-on technicians also knew, because they would have to perform the work on the payload and missile. The ultimate number was held to eighty-eight, and so participants called themselves "the 88 club." Most senior officials had no idea what their underlings were up to.

This degree of secrecy was unusual, even for a high-level defense program, but anyone who had experience with the rocket-firing community would understand why it was necessary. As much as they were warned about loose lips, workers at Canaveral, and contractor facilities like Convair's Atlas factory, sometimes had trouble keeping their mouths shut. The jobs they did were exciting and they often learned inside information on stories that became front-page news. It was hard to resist making a remark to a neighbor or friend.

The problem may have been worse among the higher-ranking officers, executives, and professionals. These men received flattering attention from reporters who worked the space beat, some of whom were becoming famous journalists. It was all too easy for an engineer to contact the friendly man from the Associated Press, or the charming woman who wrote for *Aviation Week*, to make an important point or call attention to an upcoming test. It was human nature to try to influence the press, to get positive coverage for your own programs, and to make sure that mistakes made by the "other guys" were noticed.

Altogether, the Cape and the larger missile community leaked so much information that one of SCORE's greatest achievements was that no one outside of the eighty-eight suspected what was afoot until the week before it happened. And then the only breach occurred because

General Yates needed to use the lavatory in the building at Patrick Air Force Base where Kenneth Grine had his public affairs office.

As Yates entered the men's room with another member of the 88 club, Jay Barbree, who just happened to be seated in one of the stalls, recognized his voice immediately. When Yates suggested his fellow officer check to be sure the room was empty, Barbree lifted his feet to escape being noticed. Thinking they were alone, the two men briefly discussed Missile 10-B and its prospects for going into orbit. As they left, Barbree knew he had heard an important bit of extra-fresh information.

The problem was that Barbree couldn't go on the air with a snippet of conversation overheard in the men's room. He went to Kenneth Grine for confirmation but discovered that the public affairs officer either didn't know anything about 10-B's secret purpose or he had been ordered to keep quiet. (In fact, Grine knew nothing.) Every other source Barbree contacted turned him down the same way, except for one man employed by a major missile contractor. He said the Atlas was bound for space, and even revealed that it would broadcast a message from the president. But he wouldn't allow himself to be quoted on the record. In the end, Barbree couldn't go with the story he captured while on the toilet. But at least he knew enough to approach the launch of 10-B with a sense that something special was going to happen.

★

Barbree was not alone in his suspicions. One of the launch directors for 10-B, Curtis Johnston, was not one of the 88. But he was in charge of getting the rocket off the ground, and in the weeks leading up to the launch he felt something strange was going on. The first sign was a large wooden crate, draped in a black tarp and marked "Keep Out," that arrived more than a month before the scheduled launch date. No one seemed to know what was inside the box, but it was brought inside the missile assembly building and carefully placed where it wouldn't be disturbed.

Then, as the days dwindled toward the December 18 firing, Johnston received orders to trim weight from the missile. His boss, Convair engineer and Chief Launch Director for 10-B Travis Maloy, was so serious about this job that he told Johnston to have technicians use heavy tools to remove even little bits of hardware—brackets, bolts, nuts—that were not essential. Why would weight matter so much, wondered Johnston, if this was just a test shot aimed for some spot in the Atlantic?

Two days before the launch, Maloy had a more startling assignment for Johnston. Taking him aside to speak in private, he told him to select a small number of trustworthy men to quietly replace the nose of the rocket—supposedly a package of instruments built by General Electric—with the contents of the mysterious crate in the assembly building. Maloy and Johnston were good friends and normally they would have discussed an assignment like this in detail, perhaps even sharing some inside information. This time, though, Maloy told Johnston not to bother asking questions, because he wouldn't get any answers. For the next few days Johnston should just quietly do as he was told, even if it meant breaking the rules. "Trust me," said Maloy.

At dusk, when the second shift started, Johnston took some of his men to the pad and they climbed to the top of the gantry and a platform called Deck 11. There they began the work of unbolting the stubby nose cone that had been supplied for the 10-B by GE. When this work was finished they used a crane to lift it up off the fuselage and down to the ground. In the meantime the crate had been fetched from its hiding place and brought to the pad. As it was opened Johnson found a long, hollow, tapered top—it was shaped like a dunce cap—for the Atlas. It was as empty as a beach ball.

The new nose weighed less than 200 pounds, and was easy to handle with the crane. But when they got it to the top of the gantry, Johnston found it was so long that they couldn't winch it high enough to set it down on the missile. After a little cussing and fuming, Johnston figured out a way to use ropes to lift the pointy rocket top higher while some of his men reached out from the deck to push the bottom of the cone away from the gantry. In this way they would shimmy the thing upward and then angle it down onto the Atlas. Everyone liked this idea, except the quality control inspector who was part of the team.

"I'm not going to be a witness to this," he shouted. "You don't have any paper!"

"Paper" was test range shorthand for the authorization required whenever a change was made to a missile. Sometimes the paper came from Convair headquarters in San Diego. In other cases, supervisors at the Cape were authorized to write these orders. Johnston knew he didn't have time to get paper from the West Coast, and given Travis Maloy's attitude, he was sure he didn't need it. He improvised, writing up an order and signing it himself.

Not satisfied, the inspector clambered down the gantry and went to find his supervisor. The chief inspector called senior managers in charge of the mission. Typically this kind of complaint would be resolved in the inspector's favor, because safety was a high priority. This time, however, the safety monitors received not support, but a warning. They were told that for this launch only, Johnston would be allowed to act on his own. Interference, for any reason, would be cause for dismissal.

The new nose was the subject of great speculation when the day shift arrived the next morning, but none of the questions that were asked brought real answers. Instead, those who knew about the secret plan pretended that higher-ups in San Diego and Washington were issuing ridiculous orders that they had no choice but to execute. (To make the ruse work, insiders even pretended to argue about certain commands from distant authorities, calling California and speaking so loudly that they could be heard by others.) With all that had to be done in the run-up to the launch the engineers and technicians were too busy to press the case, and so they just accepted that 10-B had a special, pointed nose cone.

When he returned to work, Johnston got another unusual request from his friend Maloy: break the connection between the missile's guidance system and a fuel cutoff switch. Without hesitating, Johnston took a technician out to the pad, where they opened up the rocket and snipped the wire. Knowing that the broken circuit would be detected in the blockhouse during the upcoming countdown, Johnston then took the job a step further. He had the man with the snippers rig a fake connection by running a wire to the blockhouse. When the time came for monitors to check the switch, everything would seem okay.

Later on this strange day, Johnston was called to meet three visitors, including an officer from the Army Signal Corps. Friendly and always at ease with people (not a typical trait among engineering types), Johnston was often summoned to give tours, and so he thought nothing of meeting the newcomers. But they didn't want a look around the launch complex. Instead, they produced a special package of instruments to be added to the missile. Again acting on Maloy's orders, Johnston quietly returned to the launch pad with this equipment and worked with technicians, who opened up the side of the Atlas and secured the Signal Corps package inside a space called the "pod." Nestled there on a small track and supported by shock-absorbing rubber pads, it would be secure for the ride.

All day long, Johnston noticed more clues that suggested this launch was different. One big tip-off was the new fueling process that would be tried with this shot. A special group of engineers had come to the Cape from a test facility in Southern California to handle this job. They put a special meter on the rocket, showing the level of liquid in the tanks. They would make sure that the missile was filled with far more fuel than was required for a simple test shot down the missile range and into the Atlantic Ocean.

With all of the unusual prelaunch happenings, Johnston wasn't the only one who developed suspicions. An Air Force propulsion expert looked at the streamlined nose, considered the missile's weight loss, and estimated that if the rocket engine burned all its extra fuel it would reach a speed of 25,000 miles per hour, more than enough to reach orbit. He wrote the number on a blackboard in an Atlas program office, challenging his bosses to explain it. They refused.

Finally, on the last night before the scheduled launch, Maloy gave Johnson the most controversial assignment of the mission. Even with the new nose, and after all the extra bits and pieces had been taken off, the "bird" was still too heavy. He told Johnson to slip out to the rocket, open the pod, and remove the missile's self-destruct mechanism. This surgery would lighten 10-B by at least eighty pounds.

Of course without this equipment, the rocket would be beyond the launch team's control the moment it left the pad. If it veered off course, headed for a nearby community, or threatened to come right back to the blockhouse, there would be nothing they could do. The Atlas would strike with all its weight, including the explosive power of whatever fuel was left in its tanks. This kind of disaster was a real possibility. Two out of the last three Atlas launches had failed. In September one had exploded just eighty seconds into flight.

Real as the danger was, Johnston didn't question the order to remove the self-destruct mechanism. By this time he was certain that 10-B was being lightened, fueled, and equipped for a mission in space. He was completely confident that Maloy had full authority for every order he issued, and that the final say-sos were probably coming from Washington. He arranged to have the equipment removed and then gave it to the chief inspector, the one who had been threatened with dismissal, for safekeeping. When the actual countdown began it was locked in a cabinet in his office.

★

On December 18, Curtis Johnston left his house in Titusville wearing a crisp white shirt, red Texas-style bow tie, a cowboy hat, and a tan sport coat. This good-luck outfit had become a kind of trademark for Johnston when he worked at the White Sands missile range in New Mexico. He first wore it to amuse his colleagues after two consecutive launches had been scrubbed for technical reasons. On that day the missile flew. The next shot, with Johnston on hand in regular clothes, was scrubbed. Ever since, he went Western on every launch day. At the Cape his colleagues, who appreciated anything that broke the tension on launch days, had encouraged him to continue the costume tradition. They had also given him a springy child's hobbyhorse. He wouldn't think of disappointing them.

At the blockhouse, where he slipped on a headset to conduct the countdown, Johnston was soon too busy to think much about 10-B's secret mission. He had to run through a list that called for dozens of procedures and checks on the missile's status. If everything on the bird worked properly and the launch went well, Johnston would discover its true purpose soon enough.

The three-hour countdown, which began a little before 3 P.M., proceeded with no unscheduled delays until a little after 5:30, when the blockhouse man assigned to range safety checked for a light indicating that the self-destruct system was working. Instead, the lights on his console signaled a problem. As the trouble was reported, Travis Maloy intervened. He called higher-ups who were in on the secret of 10-B and made a show of discussing the problem and then getting clearance to launch without the self-destruct system. On both ends of the call, onlookers were stunned by the decision. Big rockets were never launched without the protection of a self-destruct system.

The countdown continued. It reached zero at seven seconds after 6:02 P.M. Curtis Johnston pushed the button that launched Atlas 10-B and its mysterious payload. The ignition sequence—when fuel starts to flow and a missile comes to life—was critical for every launch, but knowing this rocket lacked the usual safety equipment made Johnston feel more anxious than usual. He stood up and watched with clenched fists as the ninety-foot-tall Atlas rumbled and roared upward, and he felt relief as it quickly headed out over the ocean, angling skyward and climbing fast.

The Atlas was America's first real ICBM, and in a previous test a missile from the same B series had flown more than 6,000 miles, proving it could deliver an atomic weapon from U.S. soil to the Soviet Union. Technicians tracking the flight of 10-B expected it would follow a similar path, dropping into the water somewhere in the South Atlantic. But when the main engine on the Atlas continued burning for thirteen seconds after the expected cutoff point, everyone in the blockhouse understood that 10-B was headed for a distance record. After it had been aloft for about twenty minutes, the technicians tracking its course started to worry.

"The first indication that it was going ape came from the guidance and tracking people, who told me their system had failed because the missile had gone off their scale," recalled Johnston decades later. "That's when I knew for sure that it wasn't a mistake. The bird had gone into orbit," he added.

Johnston still didn't know why his rocket had been allowed to enter space, but soon Travis Maloy invited him to find out. The two friends went to a trailer where the Army Signal Corps monitored a worldwide network of tracking and listening stations. Technicians at a station in Perth, Australia, reported that they had sent a signal to the Atlas, and that the instruments Johnston has secreted inside the pod had responded by sending out a broadcast of President Eisenhower's voice. He said:

This is the President of the United States speaking. Through the marvels of scientific advance my voice is coming to you from a

*satellite traveling in outer space. My message is a simple one.
Through these unique means I convey to you and all mankind
America's wish for peace on earth and goodwill toward men every-
where.*

Eisenhower's voice, speaking a Christmas-themed message of hope,
was the first ever sent to Earth from space. In the weeks to come, as it
made nearly 500 trips around the word, the greeting would be repeated
again and again.

But this was not the only trick the Atlas was trained to perform. Its
equipment could take instructions for playing back and erasing messages. It could also receive new voice messages and data and hold them
until they were called for by stations below. At one point Army experts at
Fort Stewart, Georgia, sent seven messages simultaneously, and the satellite was able to bounce them all back at once with perfect accuracy. In
another experiment, corpsmen in California sent a message that the
Atlas held until another group in Texas signaled for a playback. In an instant the words "This is Prado Dam, United States Army Signal Research and Development Laboratory, Corona, California" came
crackling down from space. (Atlas performed this record-store-repeat
function reliably, but not perfectly. According to one report, the tones of
a dance band broadcast by a radio station in New England triggered the
mechanism during one orbit. It took a few passes of the rocket for technicians to get the music out of the missile.)

Unfortunately for those who hoped America would get a big public
relations boost as millions worldwide tuned in to hear the Christmas
greeting, the quality of the president's broadcast was spotty, at best. The
signal was weak and only the most powerful radios could pick it up.
Most people who heard the historic message listened as it was rebroad-
cast over commercial radio. At the White House, President Eisenhower
joined reporters in his press secretary's office to hear a recording sent by
the Department of Defense. Despite heavy static and whistling in the
background, his words were audible and Ike declared the broadcast "one
of the astounding things in this age of invention."

More astounding than an old man's voice crackling through space
on a weak signal was the size of the vehicle that contained the transmit-

ter. The carcass, as Herbert York described it, weighed more than four tons. As it went into orbit America became the weightlifting champion of outer space. The United States also proved that its military, industry, and scientific community could match the command economy of the Soviet Union when it mattered.

The Atlas that speeded Ike's message into space was based on a novel design and could fly higher, faster, and farther than anything America had built before. (It was fast enough to cross the United States in fifteen minutes.) One of its key features was a pressurized fuselage made of metal skin no thicker than a 25-cent piece. Though critics like Wernher von Braun had considered the design unworkable, it was, in fact, a breakthrough. Atlas was so light that it could deliver a payload— satellite or warhead—more efficiently than any rocket with a similar thrust. In addition, the missile had a sophisticated new guidance system that allowed the ground crew to use swiveling engines to steer it into space. Previously satellites had used extra rocket motors for a final kick into orbit.

Hundreds of Atlases would be deployed as nuclear-tipped, long-range missiles. Others would serve as launchers for military and civilian space projects. Through many generations the rocket would be used to launch satellites and astronauts, including John Glenn, when he became the first American to orbit the Earth. Atlases would also launch interplanetary probes on missions to Venus, Mars, Jupiter, and Mercury. The last of the Atlases would fly in February 2005, ending a forty-seven-year history that included more than 500 tests, launches, and missions.[6]

★

Roy Johnson may have been lacking when it came to science and engineering, but the former VP from GE turned out to be right about the PR value of putting "something big up." Coming at the end of the calendar year, and near the end of America's first full year in space, the achievement dominated conversations about the state of the Cold War and stood as a highlight as the press reviewed the news of 1958. While urging vigilance, the editorial writers of The New York Times hailed "Triumphant Atlas" and noted, "the United States has not only caught up with the Soviets but has gone beyond them." This assessment was hard to refute, especially given the technology developed and employed in this new rocket.

Further reassurance soon came from Defense Secretary Neil McElroy, who used a press conference at the Pentagon to challenge the idea that the Soviets had any advantage in long-range missiles. Citing estimates agreed to by every element of the U.S. intelligence community, McElroy said the Soviets would not have a combat-ready ICBM before June, when the first Atlases were to be deployed. Overall, the United States was now in a far better position than it had been on the day *Sputnik I* went into orbit, he said.

SCORE was accompanied by good news overall for the president and the United States. At the end of 1958 it was clear that the economy had surged out of the worst recession since World War II. The stock market led the way with a 37 percent gain for the year. Income, employment, and manufacturing all rose, as did savings.

In some quarters, the *Sputnik*s were actually credited with the recovery because they forced rapid increases in government spending— about $5 billion on defense, scientific research, and space projects in the first six months alone—which "primed the pump" as John Kenneth Galbraith would say, creating jobs and making money flow. For his part, President Eisenhower saw much of his own post-*Sputnik* defense spending as a sedative to treat public anxiety rather than a wise investment in workable hardware. He estimated that two-thirds of the extra military funds went to soothe public fears rather than meet real defense needs. One way this flow of cash had accomplished this purpose was by creating jobs and easing the recession. The president had anticipated this happy result a year earlier saying, "A feeling of greater confidence in the security sphere might go over into economic confidence as well."

Other analysts would say that a growing wave of new, high-value technologies developed and manufactured in the United States was more responsible for the recovery. The semiconductor industry led this trend. In the five years since the industry's inception, sales had risen an astounding 1,200 percent, and manufacturers—Texas Instruments, Raytheon, Fairchild, Motorola—were boosting local communities with new plants that offered thousands of high-wage jobs.

Like few other inventions, the semiconductor was a breakthrough that had countless applications and started a cascade of invention, which led to integrated circuits and computer chips. The miniaturization made possible by the semiconductor aided the development of everything from nuclear weapons to portable radios. In time chips descended from the semiconductor would be produced and traded like high-volume commodities. Prices would be driven so low that they could be used to make items such as cell phones ubiquitous by the end of the century.

Of course few Americans were thinking about semiconductors as autumn turned to winter in 1958. Most were simply glad for America's economic resurgence. In the lull of the Christmas season, this rebound, joined with America's successes in space, and even the resolution of the Little Rock school integration crisis, affirmed President Eisenhower's leadership. He retired to Gettysburg for the holiday, to work on his coming State of the Union speech and dote on his grandchildren. New Year's Eve would bring a message of goodwill, similar to the one beamed to Earth by the Atlas, from Nikita Khrushchev. Lacking a satellite to broadcast it, the Soviet premier instead turned to the Mutual Broadcasting Network to say, "All of us Soviet people, from the bottom of our hearts, wish the American people well-being and happiness, a life of peace and tranquility for 1959 and the development and strengthening of friendly cooperation between our countries for the benefit of peace."

For a moment, the two sides in the Cold War military competition and the space race were essentially in equilibrium. The Soviets had put beeping balls and a dog into space. America had answered with its own satellites and a monkey. On both sides, long-range missiles were being developed but were not yet deployed. If a nuclear war broke out, mutual destruction was assured, and on this basis, peace reigned between the superpowers. Any change would require years to achieve even as the other side strained to keep pace.

The balance in space was not so firmly set. With the technology still in its infancy, big, quick strides were still possible. Having attempted three times to send a probe to the Moon, the Soviets finally succeeded in January with a rocket launched on the basis of ideas first sketched by the Chief Designer in 1955. He called the object Mechta, which meant dream, but the world would know it as *Luna I*, the first man-made object to leave Earth's gravitational influence, pass the Moon, and travel all the way to the Sun. Though *Luna I* was supposed to orbit the Moon, the Soviets would not immediately confess that they had missed their goal. Instead the mission was announced as a great achievement. In America, President Eisenhower would admit that it was, and offer congratulations. Lyndon Johnson would declare that the administration was "not going far enough" in the space contest.

By the time *Luna I* launched, the holiday season was over. Johnson's quip signaled that the conquest of space, for reasons of national pride, defense, and political gain, would remain atop the agenda in Washington. For at least a decade space would be a favorite cause for politicians, as well as an exciting frontier. Some of the most important developments would take place in absolute secrecy. The construction, launch, and successful orbit in August 1961 of the U.S. *Corona* satellite may have been the most important of these. Equipped with powerful cameras, *Corona* provided a feast of intelligence on Soviet weaponry that calmed the fears of U.S. leaders and contributed to peace.

Astounding public feats, including the landing of men on the Moon, would be realized by each side until gradually America established supremacy in both civilian and military uses of space, including reconnaissance satellites and ICBMs. Despite some close calls, including the 1962 Cuban Missile Crisis, Mutual Assured Destruction would help to prevent the hostile firing of a single atomic weapon until the Soviet Union collapsed and gave way to a new society. Remarkably, during the fifty years that followed *Sputnik,* space would never be used as a platform for war. All that began in those shaky first days of the space race, with the ball, the dog, and the monkey, would contribute to a peace that was Eisenhower's *mechta.*[7]

EPILOGUE

The competition that began with the flight of *Sputnik I* allowed America and the Soviet Union to fight a kind of bloodless war for more than three decades. Waged at a cost estimated in the trillions of dollars, it was the greatest scientific and engineering contest ever conducted. Some would argue that this battle for technological supremacy actually served as a substitute for the real thing, diffusing animosities and consuming energy and emotions that might have found more dangerous outlets. Certainly it gave the superpowers a way to exert influence, gain stature, and keep score without killing people and scarring the landscape.

Like all major conflicts, the struggle in space shaped history in both direct and indirect ways. It produced immediate human achievements—spaceflight, visits to the Moon, orbiting space stations—that ranked in importance with Europe's opening of the New World and the discoveries of atomic science. The race also contributed to innumerable advances in technologies that most wouldn't associate with space. Some were made as a direct result of the federal education programs created in *Sputnik*'s wake, which improved public schools and helped millions of students go to colleges and universities.

Most importantly, the space competition changed humanity's view of itself and its home planet. The view of Earth, as photographed from space, installed in the human mind a true image of the world as a small and precious home. This image became a symbol of a global environmental movement, by making it possible for people everywhere to see how ecosystems, weather, and even water supplies were interconnected.

Inspiring as they were, photos from space and great achievements by astronauts did not guarantee the pace of exploration. Instead, budget constraints and the ebbing of the public romance with space once the Moon was reached meant that the major programs in both the USSR and the United States waned after the 1960s. Both countries built orbiting laboratories and America developed reusable shuttles that resembled science fiction–based notions of spaceships. But manned missions to new realms were replaced by cheaper automated probes dispatched to the far reaches of the solar system. Ambitious missions were occasionally proposed—most recently some U.S. officials suggested establishing a permanent Moon base by 2024—but the cost of these endeavors tended to dominate subsequent debates. (After a total of fourteen astronauts died in two horrific shuttle disasters, the potential loss of lives was also more widely considered.)

For the individual men and women involved—politicians, astronauts, engineers, scientists, journalists—the space race offered varying risks and rewards. In rare cases, like Yuri Gagarin's first space flight and President Kennedy's commitment to land an astronaut on the Moon, events became landmarks in world history and those who participated reached a kind of immortality. ("Firsts" such as Alan Shepard's space flight for America guaranteed fame.) More commonly, people who never gained long-lasting fame nevertheless got the opportunity to be part of something grand and exciting. Among those portrayed in this book were some who counted their involvement as a moment, and others whose lives were profoundly affected over a long period of time by their experiences during the first year of the space race:

★

RICHARD RESTON, who in 1957 rode with his parents to Moscow in a Ford Fairlane, was in Edinburgh, Scotland, when *Sputnik* beeped its way into history. At the time he was fascinated by his fellow college students' stunned reaction to the satellite and struck by their deep concerns for America. Reston became a journalist and served the *Los Angeles Times* in Washington and Moscow, where he was posted less than ten years after he had visited on vacation. He spent roughly four years in the USSR, trying to practice journalism in a place and time when secrecy, propaganda, and fear reigned. One of his most vivid memories would be of the day he was suddenly summoned to see Defense Minister Rodion Malinovsky. A hero general of World War II, Malinovsky used the meeting to warn America, through Reston, that the growing war in Vietnam "will drown you."[1]

★

After suffering a heart attack in 1960, **THE CHIEF DESIGNER** was plagued by fatigue and illness. He would continue, however, to play a leading role in every aspect of the Soviet space program, from Yuri Gagarin's successful flight to the first space walk, achieved by Alexei Leonov in 1965. He planned the Soviet Union's drive toward manned Moon missions, but would not live to see this dream fulfilled.

In January 1966, the Chief Designer underwent surgery for removal of an intestinal polyp. Surgeons discovered, instead, a larger cancerous tumor and tried to remove it. Their patient survived an eight-hour surgery but died before the day was over. The death would be the subject of controversy for decades to follow, as writers debated the quality of his care.

With his demise at age sixty-one, the Soviet leadership finally decided to reveal the Chief Designer's identity and to credit him with his achievements. Sergei Pavlovich Korolev's obituary was published in *Pravda* with a lengthy accounting of his achievements and awards—including the Lenin Prize—and a photo showing him wearing his many medals.

Korolev's funeral drew cosmonauts, military leaders, scientists, engineers, government ministers, and General Secretary Leonid Brezhnev to the same Hall of Columns where Stalin had lain after death. The urn bearing his ashes was installed in the Kremlin wall. In time he would be the subject of full biographies in both the East and the West. His face would appear on a postage stamp and the Moscow-area community that was home to much of the USSR's space program would be renamed Korolev.[2]

★

JAMES VAN ALLEN appeared on the cover of *Time* magazine in 1959 as he was credited for helping America lead the world in science. The magazine honored him again at the start of 1961 when he was among twenty scientists who were named Men of the Year.

In the decades that followed *Sputnik* and his own first satellite achievements, Van Allen continued to explore space from his home base at the University of Iowa. Generations of his students participated in projects that sent instruments into Earth orbit and deep space as part of NASA programs called Pioneer, Mariner, Voyager, Galileo, and Cassini. Among his many later discoveries were a moon and radiation belts circling Saturn.

Van Allen retired in 1985, but remained an emeritus professor. He worked almost every day in a top-floor office of the Van Allen Building on the Iowa campus. His achievements earned him more than a dozen honorary doctorates and additional awards including the Gold Medal of the Royal Astronomical Society, London, and the Crafoord Prize from Royal Swedish Academy of Sciences, which among geo-scientists is regarded as a sort of Nobel Prize.

Always willing to take a stand in a polite sort of way, he became a critic of manned spaceflight, arguing that nearly all that science needed to learn about the cosmos could be discovered with machines at much lower cost and far less risk. In his eighties and nineties he often served a sort of inspirational role, meeting with students, attending conferences, and helping writers and historians piece together the story of space exploration. He spoke freely about his own quests for a life of purpose and achievement. "A lot of people were brighter and more perceptive," he said, "but I had that curiosity and I could be a bulldog when I was looking for answers." He died in August 2006, at age ninety-one, leaving his wife, Abigail, five children, seven grandchildren, countless friends, colleagues, and students, and two yet-to-be-published scientific papers.[3]

★

After stopping T. Keith Glennan's effort to take the von Braun team, **GENERAL J. BRUCE MEDARIS** tried to create an Army role in America's space future. He saw missile defense as a key mission and eagerly sought this work for his group. He also tended to the development of the Huntsville complex, monitoring every detail of life there, right down to a mosquito infestation at the golf course and the rising population of rabbits that threatened vegetation on the grounds.

Ultimately, Medaris was unable to reserve much of a space role for his service, and a second attempt to transfer much of his manpower to NASA succeeded in 1959. The general resigned from the Army in 1960 as his friend von Braun and 4,000 others in Huntsville were being transferred to the new space agency. Medaris then took the post of president of the Lionel Corporation, the manufacturer of model trains that would soon embark on a series of mergers and acquisitions to become a major supplier of electronics and other equipment to the U.S. military.

With the move to Lionel, Medaris received a salary of $60,000 a year, an enormous sum compared with a general's pay. But he also found himself immersed in constant controversy. The firm was controlled by its chairman and major stockholder, Roy Cohn, former aide to Senator Joseph McCarthy. Within a year Cohn was the subject of numerous stockholder complaints, including objections to Medaris's salary. By 1963 the company's stock, once valued at $35 a share, was worth less than $5. Cohn was ousted, and Medaris left the company shortly thereafter.

Medaris retreated to the town of Maitland, Florida, where he operated a small consulting business. New battles with cancer, which he won, helped inspire him toward religious study. He became a priest of the Episcopal Church in 1970 and preached often in churches around the country on the power of faith in healing. "All healing comes from the Lord," he told audiences.

In the years before his death in 1990, Medaris was a popular speaker at churches. He maintained his loyalty to the Army, his fascination with space, and many of his friendships with rocket scientists. He made his last appearance in the national press in 1985, when he defended Arthur Rudolph, one of the Germans brought to Huntsville after World War II, as he was accused of Nazi war crimes by the U.S. Department of Justice.

Rudolph, who was identified as a war criminal by the Army in the 1940s, gave up his American citizenship and returned to Germany, where he died in 1996. He was the only member of von Braun's team ever to be prosecuted in the United States for war crimes. [4]

★

If he had lived to the mid-1980s, when U.S. officials and journalists began new investigations into the backgrounds of scientists brought to America after the war, **WERNHER VON BRAUN,** like Arthur Rudolph, would likely have been embroiled in controversy over his Nazi past. In the decades after his death in 1977, documentaries, books, and journal articles revealed that the records of von Braun and others had been sanitized to hide their connections to Nazi atrocities. Most notable was *The Rocket and the Reich,* by historian Michael Neufeld of the Smithsonian Institution, which depicted von Braun as a patriotic German who was essentially amoral and opportunistic. He cared most about his work on new technology and personal achievement.

The revelations about von Braun and the other rocket scientists showed how the U.S. government had ignored and downplayed their Nazi activities so that they might serve as scientific assets during the Cold War. In von Braun's case, the effort put into his image allowed him to take full advantage of his opportunities with the Army and make a smooth transfer to NASA as the director of the Marshall Space Flight Center in Huntsville.

During his time at Marshall, von Braun showed a new, public concern for issues beyond rocketry and space. He became an outspoken advocate of civil rights and critic of racism, confronting his neighbors in Alabama with speeches that warned that segregation and resistance to change were hurting the state. This point of view—that racism was self-defeating—made his statements seem less like statements of conscience and more like economic advice. But voiced in the time and place where he made them, this critique showed a certain courage. It may have also been the most effective argument with most Alabamans.

On the tenth anniversary of *Sputnik* in 1967, von Braun called for extensive exploration of the Moon using the Saturn rockets, which he was developing for that purpose. These missions should be followed, he said, by voyages to Mars and greater use of satellites to aid communications, weather forecasting, agriculture, and peacekeeping on Earth.

When von Braun's *Saturn V* delivered men to the Moon in 1969, the celebration in Huntsville surpassed the one that had greeted the first U.S. satellite in 1958. "The space flights must continue," said von Braun, in the middle of a flag-waving crowd. "The ultimate destiny of man is no longer confined to Earth."

A year later, von Braun left Alabama to work for NASA in Washington, D.C. Space began to lose its luster for lawmakers and taxpayers. Faced with the expenses of the Vietnam War and soaring welfare rolls caused by a recession, President Richard Nixon steadily reduced NASA's budget. Skilled and nonskilled employees were laid off and programs were cut. A discouraged von Braun took a job in private industry. In 1975 doctors found cancer in von Braun's large intestine. In his subsequent writing and interviews he dwelt on faith and ethical issues in science.

Von Braun died in 1977 at age sixty-five. A large cultural and civic center in Huntsville was named for him, as was a scholarship at the local branch of the University of Alabama. Much honored while alive, von Braun's image was dimmed, but not completely changed, by revelations of his Nazi-era activities. He remained a controversial figure whose past motives, attitudes, and actions were open to dispute and interpretation.[5]

★

BRADFORD WHIPPLE, the twenty-year-old airman who was among the first to hear *Sputnik* and later helped steal a copy of the satellite, left Germany and the Air Force soon after the world's fair caper. The theft of the satellite would remain secret until he spoke of it decades later. (Since *Sputnik*'s components had been revealed to American scientists and the satellite contained no other secret equipment, the little spy operation yielded nothing of value.)

As Whipple left for the United States, he and Anke, still very much in love, planned to reunite. Back in Massachusetts, Whipple attended college to study Russian and German and was selected to be one of the first sixty-four members of the Peace Corps. Still corresponding with Anke, he went to Colombia for his service and asked her to join him. Though she was eager, her trip was delayed by her protective father's doubts. (With bandits controlling part of the Colombian countryside, he was concerned about her safety. It didn't help when Brad requested she bring a pistol for protection.) Believing she had rejected him, Whipple eventually married a Colombian woman. Only many years later would he learn that Anke's letter promising to come had been lost in the mails.

After working as a Peace Corps trainer, Whipple eventually returned to his native New England. He joined the faculty at Franconia College in New Hampshire. In the mid-1970s, when his father died, Whipple took over a family-run inn with cottages in the town of Sugar Hill, New Hampshire. He has lived there ever since, and served his community as, among other things, fire chief, police officer, and selectman.

Anke eventually came to live in the United States, settling with her American husband in the Midwest. She and Whipple reconnected, as friends.[6]

★

J. ALLEN HYNEK, who served as a consultant for the press in *Sputnik*'s wake and organized the nation's amateur moonwatchers, became chairman of the Astronomy Department at Northwestern University in 1960, where he taught for more than eighteen years. He also continued working as a consultant UFOs for the U.S. Air Force. Though he often found ways to explain reports of mysterious flying craft, he would not dismiss the idea that creatures from beyond Earth were visiting. "Scientists in the year 2066 may think us very naive in our denials." After ending his relationship with the Air Force in 1969, Hynek published a book called *The UFO Experience*, which challenged the government's sweeping dismissal of UFO claims. He was a technical adviser to filmmaker Steven Spielberg for *Close Encounters of the Third Kind*. The title was a term Hynek had coined for reports involving actual contact with space beings. Before he died, in 1986 at the age of seventy-five, he told a reporter that if he had made "the study of UFOs respectable," then he had made "a real contribution."[7]

★

Having done and seen just about everything there was to do and see in the space boom town of Cocoa Beach, young **ROGER DOBSON** left the family trailer park business in 1958. He attended graduate school at the University of Colorado and became a certified public accountant. From Boulder he moved to New York City to take a position at Price Waterhouse. While visiting Rochester, New York, to audit Eastman Kodak, he met and married a young schoolteacher named Marilyn Smith. In 1964 he and Marilyn settled in Cape Canaveral. He began working as a CPA and together they opened a private elementary school that grew from nine to 150 students in six years. Dobson and his partners also opened several hotels, which, given the continued growth in tourism and the space program, succeeded beyond their expectations. Friendly, successful, and well-spoken, Dobson became president of the local chamber of commerce, entered politics, and served as Brevard County commissioner from 1986 to 1990. By his seventies he was a silver-haired fixture in the community who took pleasure in imagining how his trailer park exploits in the Sputnik Era might surprise those who were sure they knew him well.[8]

★

After she spent the summer of 1958 as a student aide to the Army missile group, **JOANN HARDIN MORGAN** knew she would find her life's work in space projects. She returned to the Cape during school summers and went to work there full-time after graduating with a degree in mathematics. (Later she would earn a master's degree at Stanford.)

One of the very few women to work on early NASA programs, Morgan was the first to take a seat in one of the blockhouses where controllers conducted launches. On her first assignment a test director announced, "We don't have women here." Morgan called her boss, who instructed her to "plug in your headset and get on with your test." She did as she was told, and was able to complete the job. Rarely would she encounter similar sexism.

Morgan would participate in the Mercury and Gemini programs and serve on the launch teams for Apollo, Skylab, and Apollo-Soyuz, which brought U.S. astronauts and Soviet cosmonauts together in space. Eventually she was promoted into the management ranks, where she gained ever more responsibility. She became the first female division chief, was the first to win a NASA medal, and was the first woman senior executive at the Kennedy Space Center at Cape Canaveral. By 2002, a year before she retired, she was acting deputy director of the entire center.[9]

★

As he left his job as Mobile Mike to become a full-time space reporter for local and national radio, **JAY BARBREE** embarked on a career that would span more than fifty years. Though he would become famous covering every American space mission, he also reported on other major stories throughout the South, including the civil rights movement. Barbree would turn down offers to transfer to bigger bureaus and even NBC headquarters, preferring to stick with one of the hottest stories of the postwar era.

In the summer of 1959, Barbree again encountered the reigning Miss Space, Jo Ann Reisinger. Although her mother was worried about their prospects, the two married in 1960. They would have three children and remain together the rest of their lives.

In the 1980s Barbree was a finalist in the competition to become the first journalist in space. However his chance to participate more directly in space exploration was lost when the plan to launch a reporter was dropped after *Challenger* shuttle disaster. As he reached his seventies, Barbree would be regarded locally as an institution, ever ready with a tale and a joke. Old-timers recalled him fondly as Mobile Mike and younger people considered his the voice of space news. He would also stand among the few who were present at the beginning of the space race and the birth of NASA, and remained involved fifty years later.[10]

★

After becoming internationally famous as the woman who lent her name to a space mouse, **WICKHAM "WICKIE" STIVERS** worked for a short time as a reporter in Florida. However, the trip to New York to appear on television had made her hunger for more adventure and excitement. Divorced from Eugene Stivers, she became a flight attendant for National Airlines. Based in New York City, she spent her working days as what she called "an airborne hash-slinger," and when off-duty used her access to free flights to travel the world. She found a lifelong match with her second husband, Donald Kitzmiller, an executive at General Electric. They settled in Florida, where, fifty years after *Sputnik*, Wickie was semi-retired to a peaceful life in a quiet home on a canal.[11]

★

After Project Argus, self-made physicist **NICHOLAS CHRISTOFILOS** continued his work at the Lawrence Livermore Laboratory in California. He considered himself a practical rather than theoretical scientist and poured nearly all his time and energy into what he called his "physics." In the 1960s, Christofilos worked intently on extremely low frequency (ELF) transmissions, which were capable of carrying signals to submerged submarines. He solved many of the problems associated with ELF, reducing the power required to generate signals and cutting down the size of the antennae systems required. Eventually an ELF station was built in Wisconsin. Christofilos, however, never saw ELF in use. He died at home, of a heart attack, in 1974.[12]

★

After Project SCORE, **HERBERT YORK** continued to work as chief of the Pentagon's research and development efforts, helping to set the nation's agenda for the Cold War arms race. He would recall, with particular relish, the bureaucratic battle he led to bring Wernher von Braun to NASA and his successful effort to limit work on impractical and outlandish proposals, including a late 1950s scheme to create an antigravity system of space propulsion.

In 1961, York went back to academia, becoming the first chancellor of the University of California at San Diego and a physics professor there. He would move between the university and government for the next twenty years, often serving as an emissary or negotiator on atomic weapons issues. His efforts earned him wide recognition and many awards, including the Fermi Award from President Bill Clinton, which is given for lifetime achievement in energy-related science.

Deeply concerned about the danger of nuclear war, York wrote extensively on arms policy, including a 1987 book, *Making Weapons, Talk-*

ing Peace. He also founded the Institute on Global Conflict and Cooperation, which has sponsored studies, conferences, and other peace projects for more than twenty-five years.

Still active in UCSD community, York believes that his base in the scientific and academic world allowed him to be more effective in government and diplomacy, where he pursued both arms control and an adequate deterrent force for America.[13]

★

The cowboy of mission control, **CURTIS JOHNSTON,** stayed at Cape Canaveral for three more years after Project SCORE, launching rockets for the Air Force and NASA and even setting up a new launch complex for the new Atlas Agena missile. After Johnston handled the first launch of the Mercury program, which involved putting up an empty satellite, his superiors determined he was, in his own words, "a little too flippant to be the test conductor for the astronauts." Undaunted, he accepted transfer to the missile center at Vandenberg Air Force Base in California. A variety of jobs, including base manager at Vandenberg, led eventually to work on cruise missiles and an assignment in Saudi Arabia. He retired in 1988, but was lured back to work in 1999 by a private firm that tried to develop a reusable rocket to place satellites in orbit. None of these rockets actually flew during Johnston's tenure. The company went bankrupt and Curtis Johnston the missile wrangler finally retired for good in 2002.[14]

ACKNOWLEDGMENTS

No book springs whole from a writer's keyboard and finds its way to a reader. Between inspiration and the shelf, each one must be encouraged, shaped, and even loved a bit by a community of supporters. In this case I must count first many at Simon & Schuster, including David Rosenthal, Bob Bender, Victoria Meyer, Johanna Li, Fred Chase, and Michael Accordino. Thank you for backing me, and backing me up.

My research began with a spin around the Cape region of Florida with Lori C. Walters, a surefooted guide who is also a historian at the University of Central Florida. We met at the incomparable Alma Clyde Field Library in Cocoa, Florida, where members of the Mosquito Beaters group, led by George "Speedy" Harrell, generously shared memories, documents, and introductions to other helpful citizens, including Roger Dobson.

Archivists can be a book's best friend and this one was blessed by many. Scott Blackwell of the Florida Institute of Technology and Debra Wynne, archivist of the Florida State Historical Society, were generous with advice and recommendations. I was also aided by the staff at the National Archives, the library of Stony Brook University, and the Smithsonian Institution. And, of course, I am indebted to writers who went before me and shared both their works and their personal insights. Among them are M. G. Lord, Matt Bille, Asif Siddiqi, Roger Launius, and Michael Neufeld. Other writers who were spared phone calls and meetings are honored in my footnotes.

Naturally, those who provided me with eyewitness accounts of history are responsible for the more lively elements of my account. I owe a debt to James Van Allen, who gave me two days of his time when, as it turned out, there were precious few left. His student and friend Carl McIlwaine shared stories and insights gathered throughout his remarkable life and introduced me to his unforgettable Bedlington terrier. Likewise, the renowned physicist, writer, and wise man Herbert York told me much of what he recalled from his service, at a very high level, to President Eisenhower and the nation.

Among the many who also deserve my gratitude, are Jay and Jo Ann Barbree, Richard Reston, Bradford Whipple, Roger Dobson, JoAnn Hardin Morgan, LaVerne Hardin, Wickham "Wickie" (Stivers) Wilson, Norman Barr Jr., Curtis Johnston, Kenneth Grine, Robert Adams, David Armbrust, Stephen Wilson, William Martin, John Neilon, Thomas O'Malley, David Simons, and Herbert Gold. Others, including Dale Holzen, J. Bruce Medaris Jr., and Nancy Yasecko, provided valuable help, although their own stories do not appear.

Finally, as always, my work has been made possible by my agent, David McCormick, and by the three women who help guide my life, Toni, Elizabeth, and Amy. And I offer appreciation for Gordo, otherwise known as Old Reliable, whose contribution to America's early adventure in space was a genuine inspiration.

NOTES

PRELUDE:
FORD FAIRLANES, ATOM BOMBS, AND SATELLITES

1. James Reston, "Auto Trip to Moscow Is a Lonely Adventure," *New York Times*, Oct. 1, 1957, p. 1. James and Sally Reston, "We Motored to Moscow," *Saturday Evening Post*, March 1, 1958, p. 29. Also from author interview with Richard Reston.

2. For Cold War stalemate, paranoia, LeMay, and Eisenhower, see James Carroll, *House of War* (Boston: Houghton Mifflin, 2006), pp. 14–15, 49, 151; and John Lewis Gaddis, *The Cold War: A New History* (New York: Penguin, 2005), p. 26. Megatonnage in David Halberstam, *The Fifties* (New York: Villard, 1993), p. 29. "NSC 68: United States Objectives and Programs for National Security: A Report to the President Pursuant to the President's Directive of January 31, 1950." For bombers never built, see advertisement in *Time*, Dec. 15, 1957, pp. 77–78. "Army Missile 63 Feet Long Put on Display in Grand Central Terminal," *New York Times*, July 8, 1957, p. 1. Alsop rebuffed, see Halberstam, *The Fifties* p. 705. Stewart Alsop, "Behind Khrushchev's Smile," *Saturday Evening Post*, Feb. 2, 1958, p. 23. For more on the Alsops, see Robert Merry, *Taking on the World: Joseph and Stewart Alsop, Guardians of the American Century* (New York: Viking, 1996), p. 315.

LeMay's Pax Atomica is described in William Burrows, *This New Ocean* (New York: Random House, 1998), p. 153. Author interview with Richard Reston. Nevil Shute, *On the Beach* (New York: Morrow 1957).

3. Bruce Kunz, "1957–59 Ford Skyliner a Mechanical Marvel," *St. Louis Post-Dispatch*, Aug. 8, 2005. Author interview with Richard Reston. See also James and Sally Reston, "We Motored to Moscow," p. 29.

4. Asif Siddiqi, *Sputnik and the Soviet Space Challenge* (Gainesville: University Press of Florida, 2000), pp. 165–66.

5. Siddiqi, *Sputnik and the Soviet Space Challenge*, pp. 12–14. Further information on purges in James Harford, *Korolev* (New York: John Wiley & Sons, 1997), pp. 49–53. See also official Web site of administration of the city of Korolev: http://www.korolev.ru/.

6. For Korolev's rehabilitation, see Harford, *Korolev*, p. 91. For his personality, see Siddiqi, *Sputnik and the Soviet Space Challenge*, pp. 40–41, 512–16; quote is on p. 16; crumb cleaning on p. 118.

7. "The Wrong Stuff—A Catalogue of Launch Vehicle Failures," www.astronautics.com. Robert Godwin, editor, *Rocket and Space Corporation Energia: The Legacy of S. P. Korolev* (Burlington, Ont.: Apogee, 2001), pp. 8, 33–35. American defense officials and their response to Soviet long-range missiles in Richard Wilkins, *The Challenge of the Sputniks* (New York: Doubleday, 1958), pp. 27–31. Dulles news conference, *New York Times*, Aug. 28, 1957, p. 6.

8. Siddiqi, *Sputnik and the Soviet Space Challenge* For Reception Day, see Harford, *Korolev* p. 94.

9. Design and manufacturing of *Sputnik* in Harford, *Korolev*, pp. 127–28. Also see Siddiqi, *Sputnik and the Soviet Space Challenge*, pp. 163–65, where launch decision is noted. The legend of the mis-typed notice about the satellite conference is told in Burrows, *This New Ocean*, pp. 184–85.

10. The development of the cosmodrome, as well as details about the history of Tyuratam and Baykonur, in Siddiqi, *Sputnik and the Soviet Space Challenge*, pp. 135–38, 284. See also http://www.russianspaceweb.com/baikonur_town.html; http://www.centennialofflight.gov/essay/Dictionary/BAIKONUR/DI171.htm, U.S. Centennial of Flight Commemoration. Milan Ilnyckyj,

"The Space Race as 'Primitive' Warfare," *UBC Journal of International Affairs* (2005), pp. 19–28. Http://www.sindark.com/2006/01/contemplating-future. html. www.russianspaceweb.com/baikonur_secrecy.html.

11. Siddiqi, *Sputnik and the Soviet Space Challenge*, pp. 154–67. See also Harford, *Korolev*, pp. 3–7, 124–31. Pissing story recounted by Carl McIlwaine in interview, citing an eyewitness account: www.globalsecurity.org/wmd/world/russia/r-7.htm; and Godwin, editor, *Rocket and Space Corporation Energia: The Legacy of S. P. Korolev*.

12. "Road to the stars" quote and scene at Baykonur in Matt Bille and Erika Lishock, *The First Space Race* (College Station: Texas A&M University Press), 2004, pp. 100–101. Khrushchev's muted reaction in Burrows, *This New Ocean*, p. 194. See also William Taubman, *Khrushchev: The Man and His Era* (New York: Norton, 2000), pp. 378–79; Harford, *Korolev*, p. 121; Siddiqi, *Sputnik and the Soviet Space Challenge*, p. 169. Tass report, *Pravda*, Moscow, Oct. 5, 1957, p. 1. Author interview with Richard Reston.

CHAPTER ONE:
THOSE DAMN BASTARDS!

1. Author interviews with James Van Allen, David Armbrust, Stephen Wilson, and Carl McIlwaine. See also Carl E. McIlwaine, "Music and the Magnetosphere," *History of Geophysics*, Vol. 7 (Washington, D.C.: American Geophysical Union, 1977), pp. 129–42. Oral History of Stephen Wilson by Brian Shoemaker, Aug. 13, 2001, held by Byrd Polar Research Center of The Ohio State University, Columbus, Ohio.

2. Author interview with James Van Allen. See also James Van Allen, "What Is a Space Scientist? An Autobiographical Example, "*Annual Review of Earth and Planetary Sciences*, Vol. 18 (May 1990), James Van Allen, *Origins of Magnetosphere Physics* pp. 1–27. (Washington, D.C.: Smithsonian Institution Press, 1983), pp. 26–28. Mal Holcolm, "The Eagle and the Turtle," *Wings Magazine*, Feb. 1980. For Darwin, see http://www.bbc.co.uk/history/historic_figures/darwin_charles.shtml.

3. George Ludwig, "The First Explorer Satellites," address at University of Iowa, Oct. 9, 2004. Van Allen, "What Is a Space Scientist?"

4. IGY detail from author interview with James Van Allen as well as from

Walter Sullivan, *Assault on the Unknown* (New York: McGraw Hill, 1961), pp. 1–4, 77–88. For Stuhlinger and Van Allen, see Ludwig "The First Explorer Satellites." Meetings to discuss rocket choice from William Pickering, California Institute of Technology oral history project, http://resolver.caltech.edu/CaltechOH:OH_pirckering_2. Historian Michael Neufeld's thorough account and analysis of the rocket selection process, "Orbiter, Overflights, and the First Satellite: New Light on the Vanguard Decision," appears in Roger Launius, editor, *Reconsidering Sputnik: Forty Years Since the Soviet Satellite* (New York: Routledge, 2000). For Vanguard and Eisenhower, see Roger Launius, *NASA: A History of the U.S. Civil Space Program* (Malabar, Fl.: Krieger, 1994), p. 23. Vanguard is also reviewed in many chapters in Sullivan, *Assault on the Unknown*, and in Frederick Ordway and Mitchell Sharpe, *The Rocket Team* (Burlington, Ont.: Apogee, 2003), pp. 257–60. Public reaction to Vanguard in Kurt Stehling, *Project Vanguard* (Garden City: Doubleday, 1961), pp. 98–99; the confidential doorstop is on p. 96. For von Braun's attitudes and personality, see Bob Ward, *Dr. Space* (Annapolis: Naval Institute Press, 2005), pp. 95–110.

5. From Van Allen, *Origins of Magnetosphere Physics*, pp. 27–29, and from author interviews with James Van Allen, Stephen Wilson, and David Armbrust. See also oral history of Stephen Wilson by Brian Shoemaker, Aug. 13, 2001, held by Byrd Polar Research Center of The Ohio State University, Columbus, Ohio.

6. RCA reception in Roy Silver, "Satellite Signal Broadcast Here," *New York Times*, Oct. 5, 1957, p. 1. Author interview with Bradford Whipple. Whipple also recalled this incident in an unpublished paper, "They Were Always Turned On," 2006. Author interviews with Stephen Wilson and David Armbrust.

7. Interview with Jay Barbree. Chester B. Cunningham, "Weekly Progress Report," Lima Station, week ending Oct. 5, 1957, in files of John P. Hagen and Project Vanguard, National Archives, College Park, Maryland. The phrase "twilight zone" in "Rocket Sighting Here Not Expected for Month," *New York Times*, Oct. 25, 1957, p. 13.

8. CIA briefing and mood of the administration recalled in author interview with Herbert York, who was then chief scientist of the Pentagon's Advanced Research Project Agency. Eisenhower from Ward, *Dr. Space*, p. 112. Reaction to *Sputnik* and anecdote about the briefing at the National Academy of Sciences in Sullivan, *Assault on the Unknown*, pp. 66–72.

9. Medaris tells his life story in John B. Medaris, *Countdown for Decision* (New York: G.P. Putnam & Sons, 1960). The anecdote about the V-2 is in Gordon Harris, *A New Command* (Plainfield, N.J.: Logos International, 1976), pp. 106–7. Medaris quote about his being right in *People*, July 14, 1975, p. 44. Medaris's views on rockets and the Army were outlined in his address to the Association of the U.S. Army, Oct. 26, 1956, text in Medaris archive, Florida Institute of Technology Library, Melbourne, Florida. For Gavin, see James Gavin, *War and Peace in the Space Age* (New York: Harper & Brothers, 1958), pp. 29–30.

10. Von Braun's appearance described by Willy Ley in Dennis Piszkiewicz, *The Nazi Rocketeers* (Westport, Ct.: Praeger, 1995), pp. 23–24. The quote about how the Germans chose America in Dwayne Day, "Paradigm Lost," *Space Policy*, August 1995, p. 135. The transfer of the Germans to the United States was announced by the War Department public relations branch in a news release, "Outstanding German Scientists Being Brought to the U.S.," Oct. 1, 1945. For a complete biography of von Braun, see Ward, *Dr. Space*; the quotations from von Braun and Medaris are on pp. 99–111; the story of von Braun's arrest is on pp. 39–40; his escape to America is on pp. 54–59; his marriage to Maria von Quisdorp is on pp. 68–69. Michael Neufeld, "Wernher von Braun, the SS, and Concentration Camp Labor: Questions of Moral, Political, and Criminal Responsibility," *German Studies Review*, Vol. 25, No. 1 (2002).

11. The *Collier's* series on space was published sporadically from 1952 to 1954. *Man Will Conquer Space Soon* was published in *Collier's* March 22, 1952. The books that von Braun contributed to were *Across the Space Frontier, The Conquest of Space,* and *The Exploration of Mars.*

12. The mood at Redstone recalled in author interview with Donald Kitzmiller, who worked for Army Ballistic Missile Agency at the time. Gavin's background is covered in Stanley Falk, "Individualism and Military Leadership," *Air University Review*, July–Aug. 1980). http://www.airpower.maxwell. af.mil/airchronicles/aureview/1980/jul-aug/falk.html#falk. Gavin's relationship with Gellhorn in Brenda Maddox, "Scooping Hemingway," *New York Times*, Oct. 26, 2000, p. B30. Gavin's view of a coming Earth war in Gavin, *War and Peace in the Space Age*, pp. 208–10. The Medaris story is told in his autobiography, *Countdown for Decision*, and in Harris, *A New Command*. For details on the night of *Sputnik*, see Medaris, *Countdown for Decision*, pp. 154–64.

CHAPTER TWO:
NEW MOON WORRIES

1. Poem in Roger Launius, *NASA: A History of the U.S. Civil Space Program* (Malabar, Fl: Krieger, 1994), p. 25. Austin Wehrwein, "The Middle Class Mocked as Bland," *New York Times*, Dec. 28, 1957, p. 4. Divorce in Stephanie Coontz, "A Pop Quiz on Marriage," *New York Times*, Feb. 14, 2006, p. 12.

2. For McCarthy's effect on politics, see the Internet document, Roger Launius, *Sputnik and the Origins of the Space Age*, NASA history office, www .hq.nasa.gov/office/pao/History/sputnik/sputorig.html. Http://www.eisenhower memorial.org/presidential-papers/second-term/chronology/1957–10.htm. McCarthy's use of "sputnik" in James Carroll, *House of War* (Boston: Houghton Mifflin, 2006), p. 224.

3. "Hunk of iron almost anybody could launch," Admiral Rawson Bennett, chief of naval operations, quoted in "Satellite Belittled," *New York Times*, Oct. 5, 1957, p. 3. Soviet ruses in W. Bracewell, "Intentions and Capabilities," *Journal of Europe-Asia Studies*, Vol. 49, No. 7, p. 1350. (September 1997), "Numbers racket" quote in Carroll, *House of War*, p. 220. U-2 program development in James R. Killian, *Sputniks, Scientists and Eisenhower* (Cambridge: MIT Press, 1977), pp. 79–85. Open Skies proposal in Walt Rostow, *Open Skies*, (Austin: University of Texas Press, 1982), p. 309.

4. Interview of George E. Reedy by John Logsdon on May 20, 1992. The original transcript of the interview can be found on pp. 8–12 of "Legislative Origins of the National Aeronautics and Space Act of 1958 on NASA's Web site. The interview provides exciting insights into the origin of the U.S.-Soviet space race, and Lyndon Johnson's role. Johnson walking ranch in Lyndon Baines Johnson, *The Vantage Point* (New York: Holt, Rinehart & Winston, 1971), pp. 271–73. Oral History transcript, Eileen Galloway interviewed by Rebecca Wrights, Aug. 7, 2000, NASA Oral History Project.

5. "U.S. Stupidity Seen," *New York Times*, Oct. 6, 1957, p. 42. "Senators Attack Missile Funds Cut," *New York Times*, Oct. 6, 1957, p. 1. Leonid Sedov's comments from Paul H. Satterfield and David S. Akens, monograph, *Army Ordnance Satellite Program*, held by Army Ballistic Missile Agency, Huntsville, Alabama, 1958. Richard Witkin, "U.S. Delay Draws Scientists' Ire," *New York Times*, Oct. 5, 1957, p. 3. "Sphere Reported Sending Out Code," *New York Times*, Oct. 6, 1957, p. 42. Associated Press, "Reply from

Satellite Recorded in Australia," *New York Times*, Oct. 9, 1957, p. 12. Little's sighting in "Satellite Seen in Alaska," *New York Times*, Oct. 7, 1957, p. 1.

6. "Three Hats Among the Stars," *New York Times*, Oct. 9, 1957, p. 12. John O. Fenton, "Scientists Seek to See Satellite," *New York Times*, Oct. 9, 1957, p. 12. For Hynek, see "Four Sightings Reported," *New York Times*, Oct. 5, 1957, p. 3, and "First U.S. Glimpse," *New York Times*, Oct. 7, 1957, p. 1. Walter Sullivan, "Amateurs to Aid Satellite Hunt," *New York Times*, Sept. 3, 1956, p. 29. Milton Honig, "Jersey Home Used as IGY Center," *New York Times*, Feb. 7, 1958, p. 21. "Huge Sunspots Visible," *New York Times*, May 17, 1951, p. 23. Robert Plumb, "Air Force Sharing Big Eclipse Job," *New York Times*, May 4, 1954, p. 41. "Why Stars Twinkle," *New York Times*, July 25, 1954, p. E9. For UFOs see Associated Press, "Flying Objects Near Washington Spotted by Both Pilots and Radar," *New York Times*, July 22, 1952, p. 27. "This is the satellite," reported by United Press from Washington, D.C., in "Hoax Messages Bring Warning from FCC," *New York Times*, Oct. 7, 1957, p. 17.

7. For intelligence analysis failure, see, as an example, Harry Schwartz, "Intelligence and Moon," *New York Times*, Oct. 7, 1957, p. 16. World press in "World Newspapers See Soviet Taking Lead from United States," *New York Times*, Oct. 7, 1957, p. 17. Pilot's comments to the Chief Designer in Asif Siddiqi, *Sputnik and the Soviet Space Challenge* (Gainesville: University Press of Florida, 2000), p. 168.

8. Khrushchev and the U-2 in Philip Taubman, *Secret Empire* (New York: Simon & Schuster, 2003), p. 185. Khrushchev and the Chief Designer's activities after the launch in James Harford, *Korolev* (New York: John Wiley & Sons, 1997), pp. 126–32; and Siddiqi, *Sputnik and the Soviet Space Challenge*, "Mao Accepts Bid to Visit Moscow," *New York Times*, Oct. 27, 1957, p. 14. For Soviet dogs in space, see William Burrows, *This New Ocean* (New York: Random House, 1998), 197–99; and Siddiqi, *Sputnik and the Soviet Space Challenge*, pp. 172–75. Steven Dick and Steve Garber, *History of Research in Space Biology and Biodynamics*, U.S. Air Force Missile Development Center, Holloman Air Force Base, New Mexico, 1946–1958. Review of injuries to chimpanzees, monkeys, and man in S. B. Sells et al., *Human Factors in Space Travel* (New York: Ronald Press, 1961), p. 39. Daniel Cooley, "Portrait of the Ideal Space Man," *New York Times Magazine*, May 23, 1958, p. 24. For Stapp as hero and role model, see Jascha Hoffman, "Crash Course," *Boston Globe*, Aug. 8, 2004. For hero status of test subjects, see also Alvin Shuster, "Why Human Guinea Pigs Volunteer," *New York Times Magazine*, April 13, 1958, p. 62. Simons reported his experience in Neal Burns et al.,

Unusual Environments and Human Behaviors (London: Free Press of Glencoe, 1963), pp. 132–55. Associated Press, "Junior Chamber Bestows Honors," *New York Times*, Jan. 17, 1958, p. 13. Loyd Swenson, James Grimwood, and Charles Alexander, *This New Ocean* (Washington, D.C.: National Aeronautics and Space Administration, 1966), p. 51. Donald Janson, "Balloonist Soars to 19-Mile Record," *New York Times*, Aug. 20, 1957, p. 29. Simon's experience recounted in Donald Janson, "Balloonist Lands Safely," *New York Times*, Aug. 21, 1957, p. 1, 24 Much detail and film of Simons's flights can be seen in *History Undercover: To the Edge of Space: Project Manhigh*, History Channel, 2001. See also Gary Borkowski et al., "Laboratory Animals in Space," *Animal Welfare Newsletter*, Vol. 6, Nos. 2–4 (Winter 1995–1996).

9. For Khrushchev, see William Taubman, *Khrushchev: The Man and His Era* (New York: Norton, 2003), p. 278. For Khrushchev's reaction, see also Harford, *Korolev*, pp. 121–38. For Soviet dogs in space, see Burrows, *This New Ocean*, pp. 197–99; and Siddiqi, *Sputnik and the Soviet Space Challenge*, pp. 172–75. Khrushchev interview transcript, *New York Times*, Oct. 10, 1957, p. 10. James Reston's experience from James B. Reston, *Deadline* (New York: Random House, 1991), pp. 212–13. For James Reston's concern about pre-emptive nuclear war, see John Stacks, *Scotty* (New York: Little, Brown, 2003), p. 153. Eisenhower news conference from transcript in *New York Times*, Oct. 10, 1957, p. 14. Public opinion polls in *American Reactions to Crisis*, U.S. President's Committee on Information Activities Abroad, 1959–61, Box 5, A-83-10, Dwight D. Eisenhower Library, Abilene, Kansas. William Furlong, "Communiqué from Keokuk," *New York Times*, Sept. 29, 1957, p. 212. Transcript of Eric Sevareid's commentary in Richard Witkin, *The Challenge of the Sputniks* (New York: Doubleday, 1958), p. 7. Details on Eisenhower order regarding Redstone and Vanguard program from Constance M. Green and Milton Lomask, *Vanguard: A History* (Washington, D.C.: National Aeronautics and Space Administration, 1970).

10. Descriptions of Hagen from author interview with James Van Allen. See also Kurt Stehling, *Project Vanguard* (Garden City: Doubleday, 1961), p. 70. Hagen's notes for the Senate subcommittee are found in the National Archives box of historical documents labeled with his name. Vanguard's troubles in the summer of 1957 and details on Mazur in Stehling, *Project Vanguard*, pp. 103–24. Vanguard funding problems in Green and Lomask, *Vanguard*, as well as from Walter Sullivan, *Assault on the Unknown* (New York: McGraw Hill, 1961), pp. 84–89. Van Allen from author interview with James Van Allen.

11. For *Deal*, see Al Hibbs, "On the Edge of Space," *Engineering and Science*, Vol. 66, No. 2 (2003), pp. 37–38. For satellite swarm and nuclear airplane, see Herbert York, *Race to Oblivion* (New York: Simon & Schuster, 1970), pp 130–32. For antimissile measures, see Matt Bille and Erika Lishock, *The First Space Race* (College Station: Texas A&M University Press, 2004), pp. 140–41. Jack Beal and N. C. Christofilos, "The Astron Nuclear Accelerator," *Annual of IEEE*, 1969. For Christofilos, see also http://www.pbs.org/wgbh/amex/bomb/filmmore/reference/interview/rhodes12.html and John Finney, "Atom Inventor Was Held Crank," *New York Times*, Feb. 14, 1958, p. 1. For Corona, see Taubman, *Secret Empire*, pp. 245–70. Author interview with William Martin, early Huntsville space program worker.

CHAPTER THREE:
THE CAPE

1. Author interviews with Roger Dobson, Jack King, formerly Associated Press correspondent for Cape Canaveral, 1958, Jay Barbree, and others who lived in the area at the time. Population figure for Brevard County, Florida, from U.S. Census reports. Revenues for missile industry and fishing from Milton Bracker, "Missile Program Is Hard on Fish," *New York Times*, Dec. 15, 1957, p. 64. For motels and occupancy, see "Life in Missleland," *Time*, July 15, 1957, p. 16.

2. Author interview with LaVerne Hardin and JoAnn Hardin Morgan.

3. Author interview with LaVerne Hardin. See also John B. Medaris, *Countdown for Decision* (New York: G.P. Putnam & Sons, 1960), pp. 57–61, 134–35. Eisenhower's opinion of Medaris from author interview with Herbert York.

4. Commuting on the Cape in John Manning, "Traveling A1A to Cape Missile Center," *Memory Book* (Cocoa, FL: Central Brevard Mosquito Beaters, Alma Clyde Field Library, 2003).

5. John T. Manning, "Cape Workers Married to Their Jobs," *Memory Book*. John T. Manning, "Mosquitoes Bit First and Died Later," *Memory Book*. Memories of missile launchings from author interviews with LaVerne Hardin Morgan, fire chief Norris Gray, and John Neilon. Car races in Constance M. Green and Milton Lomask, *Vanguard: A History* (Washington, D.C.: National Aeronautics and Space Administration, 1970). Hunting and fishing sto-

ries from Robert Adams, "A Missileman's Tales from Yesteryear," *Memory Book*, and author interview with Robert Adams.

6. Recollections of Jaycee Beach from author interview with JoAnn Hardin Morgan. For a complete list of explosions and other malfunctions at the Cape, see "The Wrong Stuff" at astronautix.com. Bracker, "Missile Program Is Hard on Fish," p. 64.

7. The Vero Beach crash and others in Bob Adams, "More Tales of Early Missile Days," *Memory Book*.

8. Choice of the Cape and safety issues in John Neilon, "Eastern Launch Facilities," *Kennedy Space Center Encyclopedia of Space Science and Technology* (New York: John Wiley & Sons, 2003). See also Keith Lethbridge, "The History of Cape Canaveral: The Missile Range Takes Shape," at Spaceline .com.

9. For Bomarc, see Adams, "More Tales of Early Missile Days." J. P. Anderson, "The Day They Lost the Snark," *Air Force Magazine*, Dec. 2004.

10. Missile crash and soldering problem from Medaris, *Countdown for Decision*, pp. 122–23. Danger of explosions from author interviews with LaVerne Hardin and Norris Gray. Thor loss in "Test Missile Explodes," *New York Times*, Oct. 4, 1957, p. 11. "Missile Fired in Florida, Observers Call It a Success," *New York Times*, Oct. 12, 1957, p. 1.

11. "Ill-fated" and other details from author interview with John Neilon. For Vanguard status at Cape Canaveral, see Green and Lomask, *Vanguard*.

12. Competition and conflict between the missile programs in Trevor Gardner, "Why We Lag," in Richard Witkin, ed., *The Challenge of the Sputniks* (New York: Doubleday, 1958). Project Rearguard and other details on Vanguard from author interview with John Neilon, as well as from Green and Lomask, *Vanguard*. Generals' complaints in Barcelona and Medaris contact with Pickering in Medaris, *Countdown for Decision*, pp. 158–59. For Goddard, see Jeffrey Kluger, "He Launched the Space Age with a 10-ft. Rocket in a New England Cabbage Field," *Time*, March 29, 1999.

13. The development of *Sputnik II* is detailed fully in Asif Siddiqi, *Sputnik and the Soviet Space Challenge* (Gainesville: University Press of Florida, 2000), pp. 168–74. The Chief Designer's order suspending reviews in James

Harford, *Korolev* (New York: John Wiley & Sons, 1997), p. 132; details on Sputnik missions are also found on pp. 132–34. See also Matt Bille and Erika Lishock, *The First Space Race* (College Station: Texas A&M University Press, 2004), pp. 113–14. *Sputnik* design is also described in the Astronautix online encyclopedia at http://www.astronautix.com/craft/sputnik2.htm.

14. Robert K. Plumb, "What We Learned from the Sputniks," *New York Times*, Dec. 15, 1957, p. 199. Associated Press, "Moon Station Held Possible," *New York Times*, Oct. 19, 1957, p. 8. Medaris, *Countdown for Decision*, p. 162. Walter Sullivan, "1600 Watch Skies in Satellite Test," *New York Times*, May 18, 1957, p. 3.

15. Chuck Johnson, "What Was That Thing in the Sky, or Was It?," *Cocoa Tribune*, Oct. 7, 1957, p. 1. "Mysterious Lights Identified as PAF Base Aircraft," *Cocoa Tribune*, Oct. 9, 1957, p. 1. "Egg Shaped Mystery Craft Puzzles, Terrorizes Texas," *Huntsville Times*, Nov. 4, 1957, p. 1. "Flying Egg Stalls Car, Haunts Coast Guard," *Huntsville Times*, Nov. 5, 1957, p. 1. "Saucers Discounted," *New York Times*, Oct. 7, 1958, p. 37. "Air Force Refutes Flying Saucer Tales," *New York Times*, Nov. 16, 1957, p. 3. "Jung Discounts Flying Saucers, Links Unidentified Flying Objects to Savior Myth," *New York Times*, Aug. 10, 1958, p. 49. "Proposed Studies on the Implications of Space Activities for Human Affairs," report of the Committee on Science and Astronautics, U.S. House of Representatives, April 18, 1961.

CHAPTER FOUR:
AND A DOG SHALL LEAD THEM

1. "Soviet Exhibits a Dog for Moon," *New York Times*, Oct. 28, 1957, p. 12. "Dog in a Soviet Rocket Ascends to 130 Miles," *New York Times*, Oct. 17, 1957, p. 9.

2. Asif Siddiqi, *Sputnik and the Soviet Space Challenge* (Gainesville: University Press of Florida, 2000), pp. 173–75. Capsule design including lethal injection in William Burrows, *This New Ocean* (New York: Random House, 1998), p. 198. "Recovery of Dog Called Possible," *New York Times*, Nov. 4, 1957, p. 8. See also David Whitehouse, "First Dog in Space Died Within Hours," BBC News Online, Oct. 28, 2002. "Rocket Launching Tripled Heartbeat of Satellite's Dog," *New York Times*, March 26, 1958, p. 14. Cost of a launch in William Taubman, *Khrushchev: The Man and His Era* (New York: Norton, 2003), pp. 378–79. For Khrushchev's reaction, see James Harford,

Korolev (New York: John Wiley & Sons, 1997), pp. 132–33. See also Taubman, *Khrushchev*, pp. 378–79. "Khrushchev Gay at Kremlin Fete," *New York Times*, Nov. 8, 1957, p. 1. "Khrushchev Asks a Satellite Race," *New York Times*, Nov. 6, 1957, p. 1. Mao's remarks in "Russia: A Whoop and a Holler," *Time*, Dec. 2, 1957, p. 23. Gay Talese, "Many a Dog Has Made History," *New York Times*, Nov. 10, 1957, p. 36. "Dog of the Week," *Huntsville Times*, Nov. 11, 1957, p. 2.

3. Farnsworth Fowle, "Humane Societies Protest Use of Dog," *New York Times*, Nov. 4, 1957, p. 1. Whitehouse, "First Dog in Space Died Within Hours." "Space Dog to Get Monument," *New York Times*, Nov. 30, 1957, p. 7. "Air Force Kills a Test Bear," *New York Times*, Nov. 16, 1957, p. 3. "Air Force Stops a Bear to Test Safety Devices," *New York Times*, Nov. 14, 1957, p. 14.

4. Exchange with Eisenhower from author interview with Herbert York. It is also reported in Herbert York, *Making Weapons, Talking Peace* (New York: Basic Books, 1987), p. 109. Letter from Bernard Baruch to Eisenhower, Oct. 23, 1957, from Dwight D. Eisenhower Library, Abilene, Kansas. The Gaither report was noted in the press as early as Nov. 5, 1957, in Arthur Krock, "Origins and Development of the Missile Program," *New York Times*, Nov. 5, 1957, p. 3. The report itself is publicly available and is also analyzed cogently in James Carroll, *House of War* (Boston: Houghton Mifflin, 2006), p. 219–26. Ike's "near hysteria" in Dwight D. Eisenhower, *Waging Peace* (Garden City: Doubleday, 1965), p. 211; his analysis of the Gaither report is on pp. 220–21. John W. Finney, "Pentagon Seeks a Space Vehicle to Top Satellite," *New York Times*, Oct. 17, 1957, p. 1. Memorandum of Conference with the President, Oct. 20, 1957, Dwight D. Eisenhower Library Abilene, Kansas.

5. Von Braun comments at JPL in "Tough U.S. Task Seen," *New York Times*, Nov. 6, 1957, p. 13. Existing U.S. missile and rocket programs in Eugene M. Emme, *Aeronautics and Astronautics: An American Chronology of Science and Technology in the Exploration of Space, 1915–1960* (Washington, D.C.: National Aeronautics and Space Administration, 1961), pp. 49–63. Malina's story is told in full, and in context, in M. G. Lord, *Astro Turf: The Private Life of Rocket Science* (New York: Walker, 2005).

6. Johnson's comments after Pentagon meeting in John W. Finney, "President and Aides Study Reports on Soviet Feat," *New York Times*, Nov. 5, 1957, p. 1. Kennedy is quoted in "Kennedy Assails U.S. Missile Lag," *New*

York Times, Nov. 7, 1957, p. 16. Interview of George E. Reedy by John Logs-don on May 20, 1992. The original transcript of the interview can be found on pp. 8–12 of "Legislative Origins of the National Aeronautics and Space Act of 1958 on NASA's Web site. James Reston, "An Analysis of Reaction of Presidential Hopefuls to Crisis in Defense Policy," *New York Times*, Nov. 15, 1957, p. 6. Brewton and Reedy are explored further in Walter McDou-gall, *The Heavens and the Earth: A Political History of the Space Age* (New York: Basic Books, 1985), pp. 148–50. Republican attack on Truman State Department in Eisenhower White House staff notes #229, "Republican Plan Missile Statements," Dwight D. Eisenhower Presidential Library, Abilene, Kansas.

7. Transcript of Eisenhower speech in *New York Times*, Nov. 8, 1957, p. 10. Message to Medaris in John W. Finney, "Recovery of Missile Shot into Space Is Confirmed," *New York Times*, Nov. 8, p. 1.

8. Hays in "Democrats Back Science Speed-up," *New York Times*, Nov. 8, 1957, p. 11. Polls of U.S. public in George Leopold "Cold Warriors Look Back on First Sky Spy," *Electronic Engineering Times*, May 29, 1995, p. 1. John W. Finney, "Program Speeded," *New York Times*, Nov. 9 1957, p. 1. As-sociated Press, "Von Braun Says Five Years Are Needed to Catch Soviet," *New York Times*, Nov. 10, 1957, p. 1.

9. LeMay's bombing runs against Germany in Carroll, *House of War*, pp. 19–20; his role in the Berlin Airlift is on pp. 146–47. Jack Raymond, "LeMay Hope Marks Showing of Flag," *New York Times*, Nov. 14, 1957. p. 1. Max Frankel, "Soviet Issues Satellite Pictures," *New York Times*, Nov. 14, 1957, p. 1.

10. John Popham, "South Assesses Aftermath of Little Rock," *New York Times*, Sept. 29, 1957, p. 175. James Reston, "A Test for 1960," *New York Times*, Nov. 15, 1957, p. 6. Eisenhower letter to Edward E. Hazlett, November 19, 1957, Dwight D. Eisenhower Presidential Library, Abilene, Kansas. Dana Adams Schmidt, "King of Morocco Met by President," *New York Times*, Nov. 26, 1957, p. 1. W. H. Lawrence, "Eisenhower Suffers Chill," *New York Times*, Nov. 26, 1957, p. 1. W. H. Lawrence, "Tension Mounts at White House," *New York Times*, Nov. 27, 1957, p. 11. James Reston, "A New Spokesman," *New York Times*, Nov. 28, 1957, p. B1. Eisenhower recounts the stroke in Eisenhower, *Waging Peace*, pp. 227–28.

CHAPTER FIVE:
HOLD IT!

1. Van Allen recollections from interview with author. Other details, including telex quoted, from James Van Allen, *Origins of Magnetosphere Physics* (Washington, D.C.: Smithsonian Institution Press, 1983), pp. 54–60. For Medaris and Redstone activity in November and December 1957, see John B. Medaris, *Countdown for Decision* (New York: G.P. Putnam & Sons, 1960), pp. 165–79. Ernst Stuhlinger's historical monograph, encounter with Sedov is included in Paul Satterfield and David S. Akens, *Army Ordnance Satellite Program* Army Ballistic Missile Agency, Huntsville, Alabama, 1958.

2. For testimony at the Johnson hearings, see *Inquiry into Satellite and Missile Programs Hearings Before the Preparedness Investigating Subcommittee of the Committee on Armed Services of the United States Senate, 85th Congress.* See also Jack Raymond, "Doolittle Backs a General Staff to Rule Service," *New York Times*, Nov. 27, 1957, p. 1. "Knowledge Is Power," *Time*, Nov. 18, 1957; this profile on Teller also notes the Nobel Prize count, p. 21. Jack Raymond, "Teller Bids US Disperse and Reinforce Bases," *New York Times*, Nov. 26, 1957, p. 1. Allen Drury, "Missile Inquiry Will Open Today to Hear Teller," *New York Times*, Nov. 25, 1957, p. 1. Jack Raymond, "Jack McElroy Orders Thor and Jupiter in Production," *New York Times*, Nov. 28, 1957, p. 1.

3. For "vapors," see James R. Killian, *Sputniks, Scientists and Eisenhower* (Cambridge: MIT Press, 1977), p. xvii. "Secret" memos held in archive of General J. B. Medaris, Florida Institute of Technology Library, Melbourne, Florida. For analysis of Jupiter deployment, see Philip Nash, "The Other Missiles of October" (Ph.D. diss., Ohio University, 1994).

4. Author interviews with Jay Barbree, Gene McCall, Wickham Stivers Wilson. Description of Porter from "Sylvia Porter Smart in Looks and Writing," *Huntsville Times*, Oct. 15 1957, p. 1.

5. Author interview with Kenneth Grine.

6. Author interview with Kenneth Grine. See also "Jupiter Fails Test; Congressmen Watch," *New York Times*, Nov. 27, 1957, p. 1.

7. Recollections of nightlife and journalism at the Cape in this period from author interviews with Kenneth Grine, Jay Barbree, Wickham Stivers Wilson.

Nightclub acts noted in advertisements in *Cocoa Beach Tribune*, Nov. and Dec. 1957.

8. The Mike Wallace interview appears in Richard Witkin, *The Challenge of the Sputniks* (Garden City: Doubleday, 1958), pp. 47–48.

9. Details and quotes from Kurt Stehling, *Project Vanguard* (Garden City: Doubleday, 1961), pp. 146–51, 147–51. "The Absent Minded Professor," *Time*, May 21, 1956, p. 63. Milton Bracker, U.S. Delays Test of Space Rocket at Florida Base," *New York Times*, Dec. 5, 1957, p. 1. "Satellite Launching Hits Snag," *Iowa City Press-Citizen*, Dec. 4, 1957, p. 1.

10. US Secret Service report 1-16-602.111, Dec. 10, 1957, Dwight D. Eisenhower Library, Abilene, Kansas. Richard Mooney "Eisenhower at Farm for Long Weekend," *New York Times*, Dec. 6, 1957, p. 1. Also from author interview with Jerry Beswick. Jack Gould, "TV: Disney Goes to Mars; Film on Channel 7 Explores Man's Chance of Completing Trip to the Planet," *New York Times*, Dec. 5, 1957, p. 55. See also the Disney film, released on the DVD *Tomorrowland: Disney in Space and Beyond*.

11. Stehling, *Project Vanguard*, pp. 17–24. Also Milton Bracker, "Vanguard Rocket Burns on Beach," *New York Times*, Dec. 7, 1957, p. 1. See also Constance M. Green and Milton Lomask, *Vanguard: A History* (Washington, D.C.: National Aeronautics and Space Administration, 1970).

12. Cost of space coverage from Robert Hogan, "Televising the Space Age" (Master's thesis, University of Maryland, 2005). Reasoner's account appears in Harry Reasoner, *Before the Colors Fade* (New York: Knopf, 1981), pp. 37–41.

13. Author interview with Jay Barbree.

14. Stehling, *Project Vanguard*, pp. 17–24.

15. Richard Mooney, "Rocket Disappoints President," *New York Times*, Dec. 7, 1957, p. 1. "Capital Dismayed at Test's Failure," *New York Times*, Dec. 7, 1957, p. 1. "Flopnik" and "Kaputnik" and "Phutnik pictured in photo in *New York Times*, Dec. 7, 1957, p. 36.

16. York review of Vanguard failure from author interview with Herbert York. Thor flight in "Big Week Expected at Cape," *Cocoa Tribune*, Dec. 9, 1957, p. 1. Performance of Thor from Mark Cleary, "The 655th: Missile and Space

Launches Through 1970," 45th Space Wing Office of History, Patrick Air Force Base, Florida. Green and Lomask, *Vanguard.* Postmortem on TV-3 in Stehling, *Project Vanguard,* pp. 151–56, 157–59.

17. "Young Rocketeers," *Time,* Jan. 13, 1958, p. 50. Meyer Berger, "About New York," *New York Times,* March 12, 1958, p. 28. John Devlin, "It's Rockets, Not Rock 'n Roll, for Youths Who Dig for Facts," *New York Times,* Dec. 2, 1957, p. 29. "Rocket Fuel Fumes Kill Boy," *New York Times,* Jan. 8, 1958, p. 7. Arnold Gorneau, "A Project in Space Travel," *School Review,* Spring 1958, pp. 102–9.

CHAPTER SIX:
THE ACID TEST

1. Richard Witkin, "Rocket Men Urge U.S. Space Agency," *New York Times,* Dec. 5, 1957, p. 5. *Hearings Before the Preparedness Investigating Subcommittee of the Committee on Armed Services of the United States Senate,* 85th Congress. John W. Finney, "Army Unveils 80-Inch Satellite," *New York Times,* Dec. 12, 1957, p. 1. John D. Morris, "Bigger Rocket an Urgent Need Von Braun and Medaris Warn," *New York Times,* Dec. 15, 1957, p. 1.

2. "Random Notes in Washington: Talk of Illness Stings President," *New York Times,* Dec. 16, 1957, p. 18. Edwin Dale, "President Flying to Paris Meeting of NATO Council," *New York Times,* Dec. 14, 1957, p. 1. "NATO: We Aim to Parley," *Time,* Dec. 30, 1957. W. H. Lawrence, "Eisenhower Tired, Misses NATO Fete," *New York Times,* Dec. 17, 1957, p. 1.

3. "Paris and Return," *New York Times,* Dec. 22, 1957, p. 96. W. H. Lawrence, "Eisenhower Sees Peace More Sure," *New York Times,* Dec. 20, 1957, p. 1.

4. John W. Finney, "Soviet Said to Plan Atom Plane 'Soon,'" *New York Times,* Oct. 22, 1957, p. 1. A review of the atomic plane proposal appears in Herbert York, *Race to Oblivion* (New York: Simon & Schuster, 1972), pp. 64–74. Richard Witkin, "Pentagon to Fix Atom Plane Goal," *New York Times,* Dec. 20, 1957, p. 1. "Soviet Scientist List's future Atomic Projects," *New York Times,* Dec. 30, 1957, p. 16. Homer Bigart, "Basic Research in U.S. Is Spurred by Soviet Gains," *New York Times,* Dec. 16, 1957, p. 1. Bess Furman, "$4.6 Billion in School Aid by U.S. Is Urged as the Annual Need," *New York Times,* Dec. 21, 1957, p. 1. William White, "Fulbright Fears Disaster for U.S.," *New York Times,* Jan. 24, 1958, p. 1. Gavin's resignation was covered

closely in the press: see Alvin Shuster, "Gen. Gavin, Missile Aide, to Quit," *New York Times*, Jan. 5, 1958, p. 1; "Brucker Promises Gavin 4-Star Rank," *New York Times*, Jan. 8, 1958, p. 1; Jack Raymond, "Gavin Firm on Retirement," *New York Times*, Jan. 9, 1958, p. 1. "Scientific Study Not Necessary, Educator Says," *New York Times*, Dec. 2, 1957, p. 42.

5. Clare Boothe Luce's remarks at the Al Smith dinner in Richard Witkin, *The Challenge of the Sputniks* (Garden City: Doubleday, 1958), p. 17. "Man of the Year," *Time*, Jan. 6, 1958, p. 16. Eisenhower's point of view from Dwight D. Eisenhower, *Waging Peace* (Garden City: Doubleday, 1965), pp. 216–20.

6. John W. Finney, "Secrecy Will Veil Satellite Launching," *New York Times*, Dec. 30, 1957, p. 1. Helicopter fly-bys from author interview with Jay Barbree. Story of Big Annie from "The Flight of Big Annie," *Time*, Dec. 30, 1957, pp 12–13. For secrecy and Juno transfer to Cape, see Gordon Harris, *A New Command* (Plainfield, N.J.: Logos International, 1976), p. 184.

7. Ben Funk, "Florida Has Most Disastrous Winter," *Huntsville Times*, Feb. 18, 1958, p. 8. Misloaded paper from Witkin, *The Challenge of the Sputniks*, p. 56. Stehling quote from Kurt Stehling, *Project Vanguard* (New York: Doubleday, 1961), p. 159. Vanguard details from author interviews with Wickham Stivers Wilson and Norris Gray. See also Kurt Stehling, *Project Vanguard* (Garden City: Doubleday, 1961), pp. 158–78. See also Milton Bracker, "Vanguard Firing Put Off Ending 4-Day Navy Effort," *New York Times*, Jan. 27, 1958, p. 1. Disintegration of *Sputnik I* in "Science Notes," *New York Times*, Jan. 19, 1958, p. E 11. Sam Falk, "Men of the Space Age," *New York Times*, Dec. 15, 1957, p. 224.

CHAPTER SEVEN:
BEING NONCHALANT AND LIGHTING UP A MARIJUANA

1. Party planning noted in John B. Medaris, *Countdown for Decision* (New York: G.P. Putnam & Sons, 1960), p. 190. "Plowing" in *Huntsville, Alabama* by the Huntsville Industrial Expansion Committee, 1945.

2. Bureau of the Census, May 1968, Current Population Reports and Consumer Income. $100 million from "A Fantastic Payroll," *Huntsville Times*, Oct. 13, 1957, p. 4. John Morris, "House Unit Votes $1.2 Billion to Speed Missiles," *New York Times*, Jan. 22, 1958, p. 1. Details on Huntsville from

Bob Ward, *Dr. Space* (Annapolis: Naval Institute Press, 2005), pp. 75–79. "Squalid slums" in "Rocket City USA," *Time*, Feb. 17, 1958, p. 23. For a thorough review of race relations in Huntsville in the 1950s and 1960s, see the documentary by Sonnie Hereford, *A Civil Rights Journey*, Calhoun College Center for the Study of Southern Political Culture, Calhoun Community Community College Decatur, Alabama, 1996

3. "Luncheon given by Mrs. Medaris," *Huntsville Times*, Oct. 4, 1957, p. 5. "Redstone Claim Backed," *Huntsville Times*, Oct. 8, 1957, p. 4. "Jupiter C Fever Spreads to All Corners of Nation," *Huntsville Times*, Dec. 11, 1957, p. 1. "Von Braun named for Citizen Award," *Huntsville Times*, Dec. 12, 1957, p. 1.

4. Medaris, *Countdown for Decision*, pp. 196–98. See also Gordon Harris, *A New Command* (Plainfield, N.J.: Logos International, 1976), pp. 184–85.

5. Stehling comments in Kurt Stehling, *Project Vanguard* (Garden City: Doubleday, 1961), p. 175. For Yates, see "Canaveral Revisited," *Time*, Feb. 10, 1958, p. 43. Other information from author interviews with Kenneth Grine and Robert Moser. Also from speech by George Ludwig, "The First Explorer Satellites," presented Oct. 9, 2004, University of Iowa. See also Medaris, *Countdown for Decision*, pp. 200–209.

6. Notice of the impending launch in Milton Bracker, "Army Takes Over Satellite Firing with Jupiter-C," *New York Times*, Jan. 28, 1958, p. 1. Jack Raymond, "Soviet's Biggest Satellite Reported in Countdown," *New York Times*, Jan. 31, 1958, p. 1.

7. Karen Raley, "The Town of Indialantic Turns 50," *Indian River Journal* (Cocoa, Florida, Brevard County Historical Commission), Vol. 1, No. 1, (2002), p. 14. Author interview with Robert Moser, who was in the blockhouse for the launch, provided much of the description of this launch. Other material came from the detailed contemporary account "Tenseness, Anxiety Marked Satellite Launching Hours," by Army public information officer Gordon Harris and published in *Huntsville Times*, Feb. 3, 1958, p. 14. "Inside Rocket View," *Life*, Feb. 10, 1958, p. 13. See also Medaris, *Countdown for Decision*, pp. 210–26. Matt Bille and Erika Lishock, *The First Space Race* (College Station: Texas A&M University Press, 2004), pp. 128–33. Also Milton Bracker, "Jupiter C. Is Used," *New York Times*, Feb. 1, 1958, p. 1.

8. Brucker–Medaris exchange from Medaris, *Countdown for Decision*, p. 223. Van Allen's reaction to amateur radio enthusiasts noted in interview with

author and also James Van Allen, *Origins of Magnetosphere Physics* (Washington, D.C.: Smithsonian Institution Press, 1983), pp. 58–59. Von Braun from Ludwig, *"The First Explorer Satellites."*

9. The report of the satellite radio signals from Ludwig, "The First Explorer Satellites." For the sound of *Explorer I*, visit the Web site of the radio Amateur Satellite Corporation, http://www.amsat.org/amsat/features/sounds/firstsat. html. War Room reaction from Van Allen, *Origins of Magnetosphere Physics*, p. 59. Eisenhower's reaction in Felix Belair Jr., "Success Attained," *New York Times*, Feb. 1, 1958, p. 1; also noted in Bille and Lishock, *The First Space Race*, p. 133.

10. Eisenhower's congratulatory note and Medaris's reaction in Medaris, *Countdown for Decision*, p. 196. National Academy of Sciences press conference recalled by William Pickering in California Institute of Technology Oral History Project interview conducted December 1978 by Mary Terrall, available at http://resolver/caltech.edu/CaltechOH:OH_pickering_11. Richard Mooney, "Army's Explorer Cheers Congress," *New York Times*, Feb. 2, 1958, p. 1. Soviet response to *Explorer* in William Jordan, "Soviet Scientists Congratulate U.S. on Good News of Satellite," *New York Times*, Feb. 2, 1958, p. 1. Celebration in Huntsville in Medaris, *Countdown for Decision*, p. 226. Wilson burned in effigy from "Wilson Can't Figure Out Why They Burned Him in Effigy," *Huntsville Times*, Feb. 2, 1957, p. 1. William Burrows, *This New Ocean* (New York, Random House 1998), p. 209. "Space: Voyage of the Explorer," *Time*, Feb. 10, 1958, p. 15. Celebration and effigy also noted in Harris, *A New Command* pp. 189–93. "OGMS [Ordinance Guided Missile School] Birthday Suffers a Chill, *Huntsville Times*, Feb. 17, 1958, p. 1.

CHAPTER EIGHT:
A NEW ERA OF EXPLORATION

1. "Van Allens Invited to Dinner at White House," *Iowa City Press-Citizen*, Feb. 3, 1958, p. 1. "Eisenhower Has a Slight Cold but Attends White House Dinner," *New York Times*, Feb. 5, 1958, p. 1. "White House Guest List," *New York Times*, Feb. 5, 1958, p. 13. For more on Anna Russell, see http://annarussellshrine.tripod.com/ and her autobiography, *I'm Not Making This Up You Know* (London: Continuum, 1985).

2. Von Braun comments in "Nearly All News Goes into Orbit," *Huntsville Times*, Feb. 2, 1958, p. 1. Vanguard's innovations are discussed throughout

Kurt Stehling, *Project Vanguard* (Garden City: Doubleday, 1961), especially pp. 125–41. Medaris's fight for space work described in James R. Killian, *Sputniks, Scientists and Eisenhower* (Cambridge: MIT Press, 1977), p. 127. PSAC's work described in Herbert York, *Race to Oblivion* (New York: Simon & Schuster, 1970), pp. 114–17, as well as in Keith Glennan, *The Birth of NASA* (Honolulu: University Press of the Pacific, 1993), pp. xx–xxi. Eisenhower's affection for science and scientists and concern about military noted in Killian, *Sputnik Scientists and Eisenhower*, pp. 217–25. Text of President Eisenhower's Message to Congress on the State of the Union, *New York Times*, Jan. 10, 1958, p. 8. *Legislative Origins of the National Aeronautics and Space Act of 1958* (Washington, D.C.: NASA History Division Office, July 1998).

3. David M. Nichol, "Germany Tries SS Group Who Butchered 208," *Huntsville Times*, Feb. 5, 1958, p. 12. For von Braun and other Germans who came to America, see account of their experiences at the end of the war in Frederick Ordway and Mitchell Sharpe, *The Rocket Team* (Burlington, Ont.: Apogee, 2003), pp. 181–89. *Time* cover story on von Braun, Feb. 10, 1958. Wayne Philips, "Von Braun Played Major Role in Developing a Successful Earth Satellite," *New York Times*, Feb. 2, 1958, p. 43. "New Space Projects Cited by Von Braun," *New York Times*, Feb. 18, 1958, p. 21. Von Braun speech delivered Feb. 17, 1958, meeting of the National Military Industrial Conference, Chicago; see www.ccastronomy.or/education_alert_von-braun4.

4. "Vanguard Blows Up in Second Try," *Iowa City Press-Citizen*, Feb. 5, 1958, p. 1. Vern Haugland, "For One Fine Minute Vanguard Hopes Rose," *Huntsville Times*, Feb. 5, 1958, p. 16. Stehling, *Project Vanguard*, pp. 173–81. Milton Bracker, "Vanguard Crash Caused by Flaws in Engine System," *New York Times*, Feb. 6, 1958, p. 1.

5. Milton Bracker, "An Atlas ICBM Destroys Itself," *New York Times*, Feb. 8, 1958, p. 1. John W. Finney, "2nd U.S. Explorer Fired, Vanishes," *New York Times*, March 6, 1958, p. 1. John W. Finney, "Second Explorer Failed to Orbit, Rocket Is Blamed," *New York Times*, March 7, 1958, p. 1. "President Disappointed by Explorer II Failure," *New York Times*, March 7, 1958, p. 8.

6. Killian, *Sputniks, Scientists and Eisenhower*, pp. 222–28. "Firing of Vanguard Is Postponed Again," *New York Times*, March 8, 1958, p. 18. Richard Witkin, "Vanguard Firing Likely This Week," *New York Times*, March 10, 1958, p. 3.

7. Vanguard launch story from author interview with John Neilon; and Stehling, *Project Vanguard*, pp. 182–219. Also Walter Sullivan, *Assault on the Unknown* (New York: McGraw Hill, 1961), pp. 77–108.

8. Charles Frankel, "Third Great Revolution of Mankind," *New York Times Magazine*, Feb. 9, 1958, p. 11. "Soviet Magazine Says Senators Want U.S. Moon Bases," *New York Times*, Feb. 8, 1958, p. 8. Jack Raymond, "Air Force Urges '58 Shot at Moon," *New York Times*, Feb. 26, 1958, p. 1.

9. Richard Witkin, "Chimpanzees Pass Space Speed Test, Withstand 100 G's and More in Abrupt-Stop Rides in Ground Rocket Sled," *New York Times*, Jan. 31, 1958, p. 9. "Airman Is Now Flourishing in Space Ship," *Huntsville Times*, Feb. 10, 1958, p. 1. "Space Cabin Roomier, Dotes on Mood Music," *Huntsville Times*, Feb. 12, 1958, p. 1. "Mock Moon Trip Near Completion," *Huntsville Times*, Feb. 14, 1958, p. 1. "Travel in Space? It's Easy!," *Huntsville Times*, Feb. 17, 1958, p. 1. "Space Test Airman Arrives for Visit," *New York Times*, Feb. 21, 1958, p. 25. Neal Burns et al. *Unusual Environments and Human Behaviors* (London: Free Press of Glencoe, 1963), pp. 132–55.

10. Sullivan, *Assault on the Unknown*, p. 104. Richard Witkin, "Third U.S. Satellite Fired into Orbit," *New York Times*, March 27, 1958, p. 1. "Scientists Expect Explorer III to Give Varied Data," *New York Times*, March 27, 1958, p. 15.

11. Author interviews with James Van Allen and Carl McIlwaine. See also James Van Allen, *Origins of Magnetosphere Physics* (Washington, D.C.: Smithsonian Institution Press, 1983), pp. 64–72.

12. William Jordens, "Soviet Satellite Weighing 1.5 Tons Fired into Orbit," *New York Time*, May 16, 1958, p. 1. Russell Baker, "Symington Finds President Drifts," *New York Times*, May 30, 1959, p. 9.

CHAPTER NINE:
OPPORTUNISTS AND ADVENTURERS

1. UFO reports chart by Mark Cashman, http://www.temporaldoorway.com/ufo/wave/index.htm. Max Frankel, "Soviet Pride Led by Slogan Drive," *New York Times*, Jan. 18, 1958, p. 1. Louis Calta, "Morality Story Will Be Staged," *New York Times*, Feb. 27, 1958, p. 23. "Food 'Market Basket' at Record Cost

Level," *New York Times*, April 29, 1958, p. 20. "Consumer Spending Trends," *New York Times*, May 4, 1958, p. E10. Richard E. Mooney, "Food Costs Life Price Index to a Record Level," *New York Times*, April 24, 1958, p. 1. www. scifi.com/twilightzone/. http://www.rodserling.com/

2. Author interviews with Jay and Jo Ann Barbree and Thomas O'Malley. Milton Bracker, "Cape Canaveral Land Boom Is On," *New York Times*, Feb. 17, 1958, p. 1. Richard Witkin, "Rocket Fails in Moon Shot, Blowing Up in 77 Seconds," *New York Times*, Aug. 18, 1958, p. 1. "Polaris Blown Up Over Testing Site," *New York Times*, Sept. 25, 1958, p. 5.

3. Record of missiles in this period in Walter Sullivan, *Assault on the Unknown* (New York: McGraw Hill, 1961), pp. 99–108. "Science Notes: Intense Radiation Is Reported by Explorer Satellites," *New York Times*, May 4, 1958, p. E9. John W. Finney, "U.S. Lags in Space Race 8 Months After Sputnik I," *New York Times*, May 25, 1958, p. 1. Author interview with JoAnn Hardin Morgan.

4. Author interviews with Wickham Stivers Wilson, Jack King, and Jay Barbree. See also Milton Bracker, "A Three Engine Atlas Fails," *New York Times*, July 20, 1958, p. 1. Milton Bracker, "Third Missile Fired Across Atlantic in Re-entry Test," *New York Times*, July 24, 1958, p. 1. "Mouse in Missile Met the Press Before Taking Off into Space; Air Force Provided an Introduction and Gave Details of Housing Conditions—'Wickie' Was Named for Reporter," *New York Times*, July 24, 1958, p. 10. "Sea Search for 'Wickie' Abandoned by Air Force," *Cocoa Tribune*, July 25, 1958, p. 1. "Space Mouse Details Revealed at Patrick," *Cocoa Tribune*, July 25, 1958, p. 1. Milton Bracker, "Air Force Halts Search for Mouse," *New York Times*, July 26, 1958, p. 5.

5. Author interview with Bradford Whipple. See also http://usinfo.state.gov/journals/itps/0406/ijpe/brussels.htm; http://www.atomium.be/HTMLsite/dyn/eindex.html; http://users.skynet.be/rentfarm/expo58/index.htm. "Ad Men Briefed on Tour Abroad," *New York Times*, Jan. 21, 1958, p. 10. "U.S. Aides Eschew Pessimism on Fair," *New York Times*, Feb. 16, 1958, p. 79. Walter Waggoner, "Brussels Invites the World to Its Fair," *New York Times*, March 2, 1958, p. 6. "Rivalry in Brussels," *New York Times*, Jan. 9, 1958, p. 32. Walter Waggoner, "U.S. to Be Candid at Brussels Fair," *New York Times*, March 5, 1958, p. 4. "Fair Exhibits to Symbolize U.S. Problems; Resources, Slums and Segregation Handled in Brussels Display Pictures; Charts Will Be Used to Show Progress to Fuller Democracy," *New York Times*, March 11, 1958,

p. 12. Brendan M. Jones, "Displays of U.S. Achievements Counter Huge Soviet Spending," *New York Times*, April 13, 1958, p. F1. Walter Waggoner, "U.S. vs. USSR at Brussels Fair, Too," *New York Times*, April 27, 1958, p. E6.

6. N. C. Christofilos, "The Argus Experiment," *Proceedings of the National Academy of Sciences*, Vol. 48, No. 8 (Aug. 15, 1959), pp. 1144–52. James R. Killian, *Sputniks, Scientists and Eisenhower* (Cambridge: MIT Press, 1977), pp. 186–91. Author interview with engineer and Argus participant Robert Adams. See also Sullivan, *Assault on the Unknown*, p. 129. York quotes on Christofilos from author interview. Orion from Herbert York, *Making Weapons, Talking Peace* (New York: Basic Books, 1987), pp. 150–52.

CHAPTER TEN:
EGGHEADS AND PIE TRUCKS

1. Drew Pearson, "The Washington Merry-Go-Round," *Huntsville Times*, Feb. 2, 1958, p. 4. Jack Gould, "Rise and Fall of the Quiz Empire," *New York Times Magazine*, Sept. 28, 1958, p. 12. "Sharp Rise in Science Study," *New York Times*, Sept. 7, 1958, p. 140. Gerald Walker, "Now a Varsity Letter for Scholarship," *New York Times Magazine*, Nov. 9, 1958, p. 35. Author interview with James Van Allen.

2. For "better-dead-than- Red" amendment, see Jerry K. Sweeny, "The Better Red Than Dead Amendment," at American Diplomat.org, July 2004. http://www.unc.edu/depts/diplomat/archives_roll/2004_04-06/sweeney_russel/sweeney_russel.html. "Transcript of the President's News Conference," *New York Times*, Oct. 2, 1958, p. 18. Bess Furman, "Schools Getting U.S. Science Aid," *New York Times*, Nov. 3, 1958, p. 1. Roger Bybee, *The Sputnik Era: Why Is This Educational Reform Different from All Other Educational Reforms*, National Research Council, 1998, http://www.nas.edu/sputnik/bybee.doc. Russell Baker, "Nixon on Coast Says GOP Gains," *New York Times*, Oct. 15, 1958, p. 31.

3. "Soviet Marks Sputnik 1 Ridiculing U.S. Efforts," *New York Times*, Oct. 5, 1958 p. 15. National priorities on space from statement by the President of the United States: Science Advisory Committee: Introduction to Outer Space, White House, March 26, 1958, Dwight D. Eisenhower Library, Abilene, Kansas. "Soviet Says It Recovered Dogs Shot 281 Miles Up," *New York Times*, Aug. 30, 1958, p. 1. "Soviet Spokesman Hints at a Spaceman Program," *New York Times*, Aug. 26, 1958, p. 2.

4. Army space programs described in Helen Brents Joyner, "History of the Redstone Missile System," Army Missile Command historical monograph No. AMC 23M, Huntsville, Alabama, 1965. See also David Akens, *Army Ordnance Satellite Program*, Army Ballistic Missile Agency, Huntsville, Alabama November 1, 1958. For support for Army lunar bases, see "In My Opinion," *Missiles and Rockets*, Oct. 13, 1958, distributed by Army Ordnance Missile Command. Advantages of Moon-based weapons in Jeffrey Richelson, "Shooting for the Moon," *Bulletin of the Atomic Scientists*, Sept.–Oct. 2000, pp. 22–27. Richard Witkin, "Thor Given Lead over the Jupiter," *New York Times*, May 5, 1958, p. 1. "Missile Chief Sees Success for Atlas," *New York Times*, Sept. 22, 1958, p. 14.

5. Jack Raymond, "McElroy Reveals Studies to Rebut Soviet ICBM Lead," *New York Times*, Jan. 25, 1959, p. 1. Ramo biography available from Institute of Electrical and Electronics Engineers at www.IEEE.org. Bell X-2 Starbuster NASA fact sheet, http://www.nasa.gov/centers/dryden/news/FactSheets/FS-079-DFRC.html. Loyd Swenson, James Grimwood, and Charles Alexander *This New Ocean* (Washington, D.C.: National Aeronautics and Space Administration, 1966), pp. 90–93. "Space Medicine in Project Mercury," http://lsda.jsc.nasa.gov/books/mercury/ch02.htm. "First Up," *Air & Space*, Aug.–Sept. 2000. Author interview with Kenneth Grine. Virginia Campbell, "How RAND Invented the Postwar World," *Invention and Technology*, Summer 2004, p. 50. LeMay and the RAND report also recalled by Herbert York in author interview.

6. Author interview with Jay Barbree. See also John Shanley, "TV and the Rocket," *New York Times*, Aug. 24, 1958, p. 13, and "Networks to Televise Moonrocket Firing," *New York Times*, Oct. 11, 1958, p. 14. Air Force public relations effort in Paul Satterfield and David Akens, Historical monograph, *Army Ordnance Satellite Program*, Army Ballistic Missile Agency, Huntsville, Alabama, 1958. Richard Witkin, "U.S. Rocket Rising 80,000 Miles but Will Not Circle the Moon," *York Times*, Oct. 12, 1958, p. 1. http://www.astronautix.com/craft/pioer012.htm. Richard Witkin, "Orbit Signal Fails," *New York Times*, Oct. 13, 1958, p. 1. Richard Witkin, "Air Force Gets 2 Extra Probes; May Launch a Rocket at Venus," *New York Times*, Oct. 14, 1958, p. 1. Walter Sullivan, "Rocket Supports Magnetic Theory," *New York Times*, Oct. 14, 1958, p. 1.

7. James Harford, *Korolev* (New York: John Wiley & Sons, 1997), pp. 132–38. Asif Siddiqi, *Sputnik and the Soviet Space Challenge* (Gainesville: University Press of Florida, 2000), pp. 190–95. www.russianspaceweb.com/spacecraft_

planetary_lunar.html. http://www.friends-partners.org/partners/mwade/craft/
lunae1.htm. "Pioneer's Glorious 'Failure,'" *New York Times*, Oct. 13, 1958, p.
28. Khrushchev conceives use of silos for Soviet long-range missiles, http://
www.astronautix.com/astros/khrhchev.htm

8. Roger Bilstein, *Orders of Magnitude: A History of NACA and NASA,
1915–1990* (Washington, D.C.: National Aeronautics and Space Administration, 1989). Glennan from T. Keith Glennan, *The Birth of NASA* (Honolulu: University Press of the Pacific, 2005), pp. 4–15. Von Braun quoted
in Bob Ward, *Dr. Space* (Annapolis: Naval Institute Press, 2005), p. 120.
Michael Neufeld, "The End of the Army Space Program," *Journal of Military History*, July 2005, pp. 737–57. "New Agency Asks Army to Give Up
Scientists," *New York Times*, Oct. 15, 1958, p. 1. Medaris and Watson from
John B. Medaris, *Countdown for Decision* (New York: G.P. Putnam &
Sons, 1960), pp. 244–47. *St. Louis Globe-Democrat* editorial, Oct. 18,
1958.

9. Jack Raymond, "Threat to System Seen," *New York Times*, Oct. 16, 1958, p.
14. Jack Raymond, "Space Unit Plans an Orderly Shift," *New York Times*,
Oct. 24, 1958, p. 13. Jack Raymond, "Gavin Firm on Retirement," *New York
Times*, Jan. 9, 1958, p. 1. Best-seller list from *New York Times Book Reviews*,
Oct. 19, 1958, p. 8. Gavin's description of future wars is found throughout
War and Peace in the Space Age (New York: Harper & Brothers, 1958) but is
detailed most concisely in Chapter 7, "The Decade of Decision: 1955–1965."
"Transcript of the President's News Conference on Foreign and Domestic
Matters," *New York Times*, Aug. 28, 1958, p. 10.

CHAPTER ELEVEN:
THE MONKEY AND THE PRESIDENT

1. Eisenhower story from Herbert York, *Making Weapons, Talking Peace*
(New York: Basic Books, 1987), p. 54.

2. Jack Raymond, "Pentagon Forms a Space Agency, G.E. Aide Is Chief,"
New York Times, Feb. 8, 1958, p. 1. "Man for the Job," *Time*, Jan. 5, 1959, p.
23. "SCORE," *Time*, Dec. 29, 1958, p. 7. Ginger Carter, "Public Relations
Enters the Space Age," presented at annual meeting of Association for Education in Journalism and Mass Communication, Sept. 28, 1997. "GE '54 Sales
Off, Rise for '55 Seen," *New York Times*, Dec. 16, 1955, p. 61. Author interview with Herbert York.

3. Published statement issued by ABMA, Huntsville, Alabama, Oct. 15, 1958. Author interview with Herbert York. Richard Witkin, "Explorer V Is Fired but Fails to Orbit," *New York Times*, Aug. 25, 1958, p. 1. "Jupiter Blows Up in Air, Falls at Florida Base," *New York Times*, Oct. 10, 1958, p. 10. John W. Finney, "Vehicle Won't Reach Its Goal, Scientist Says in Washington," *New York Times*, Dec. 7, 1958, p. 1. James Van Allen, *Origins of Magnetosphere Physics* (Washington D.C.: Smithsonian Institution Press, 1983), pp. 85–100.

4. "California Sets a Space School," *New York Times*, Jan. 5, 1958, p. 84. Gladwin Hill, "Births in Space Held Possible," *New York Times*, June 11, 1958, p. 17. For *The Queen of Outer Space*, see http:www.imdb.com/title/tt0052104/ as well as *Cocoa Tribune*, Dec. 15, 1958, p. 4. John Finney, "Capsule to Carry Man in Space to Be Built by St. Louis Concern," *New York Times*, Jan. 13, 1959, p. 14.

5. Loyd Swenson, James Grimwood, and Charles Alexander, *This New Ocean* (Washington, D.C.: National Aeronautics and Space Administration, 1966), pp. 110–31; for the first astronaut's duties and experience, see pp. 352–58. Flight of Gordo from author interview with Norman L. Barr Jr., MD. See also Norman Barr Sr., "A Capsuled Monkey Blazes Trail for Mankind," *Life*, Dec. 1958, p. 22. Richard E. Mooney, "Monkey in Army's Missile Fired 300 Miles in Space," *New York Times*, Dec. 14, 1958, p. c1. "Monkey Is Lost but Jupiter Flight Is termed Success," *Cocoa Tribune*, Dec. 15, 1958, p. 1. http://www.as tronautix.com/lvs/jupiter.htm.

6. Based on author interviews with Jay Barbree, Kenneth Grine, Herbert York, and Curtis Johnston. See also "How Insiders Kept Their Great Secret," *Life*, January 5, 1959, p. 20. "SCORE," *Time*, Dec. 29, 1958, p. 7. "The Big Bird Orbits Words of Peace," *Life*, January 5, 1959, p. 10. Jack Raymond, "Pentagon Confirms Full Atlas Success in 6,325-Mile Shot," *New York Times*, Nov. 30, 1958, p. 1. Jack Raymond, "Voice from Space," *New York Times*, Dec. 20, 1958, p. I1. "Atlas in Orbit," *Time*, Dec. 29, 1959, p. 31. "Comet Seen in Pacific Said to be Dying Atlas," *New York Times*, Jan. 22, 1959, p. 3. Dance band music anecdote from Chuck Walker, *Atlas: The Ultimate Weapon* (Burlington, On.: Apogee, 2005), p. 229; Walker also details the characteristics and accomplishments of the Atlas rocket, pp. 261–64, 281–82.

7. Jack Raymond, "McElroy Reveals Studies to Rebut Soviet ICBM Lead," *New York Times*, Jan. 23, 1958, p. 1. "U.S. Said to Gain in Missiles Race,"

New York Times, Dec. 30, 1958, p. 1. Edwin Dale, "Lessons of the 1957–58 Recession," *New York Times Magazine*, Oct. 5, 1958, p. 7. "Business in 1958," *Time*, Dec. 29, 1958, p. 44. Eisenhower's views of extra defense spending laid out in Roger Payne, "Public Opinion and Foreign Threats: Eisenhower's Response to Sputnik," *Armed Forces and Society*, Vol. 21, No. 1 (Fall 1994), p. 89. "Transistor Transition," *Time*, Dec. 22, 1958, p. 60. For *Luna I*, see James Harford, *Korolev* (New York: John Wiley & Sons, 1997), pp. 139–45. John W. Finney, "Eisenhower Hails Soviet Space Shot," *New York Times*, Jan. 4, 1958, p. 1. *Corona* from Director of Central Intelligence John Deutsch address to Corona Symposium, May 1995, http://www.fas.org/irp/cia/product/dci_speech_52395.html.

EPILOGUE

1. Author interview with Richard Reston.

2. For Korolev's death, see James Harford, *Korolev* (New York: John Wiley & Sons, 1997), pp. 276–83.

3. Author interview with James Van Allen. See also Van Allen's *Time* cover photo, May 4, 1959. He was among the scientists named Men of the Year, Jan. 2, 1961. An archive of material related to Van Allen's life and death is available online by the University of Iowa; see http://itsnt166.iowa.uiowa.edu/uns-archives/2006/august/080906van-allen-death.html.

4. Medaris's life story is told in both his memoir, John B. Medaris, *Countdown to Decision* (New York: G. P. Putnam & Sons, 1960), and in Gordon Harris, *A New Command* (Plainfield, N.J.: Logos International, 1976). See also "Medaris Is Named Lionel President," *New York Times*, Aug. 4 1960, p. 31. "Lionel Discloses Backers of Cohn," *New York Times*, April 9, 1960, p. 26. "Memo to RSA Post Commander Re: Rabbits, from J. B. Medaris," May 6, 1959, Medaris archive, Florida Institute of Technology, Library, Melbourne, Florida. "Memo to RSA Post Commander Re: Mosquitoes, from J. B. Medaris," July 6, 1959, Medaris archive, Florida Institute of Technology, Melbourne, Florida. Elizabeth Fowler, "Agenda Superceded by a Noisy Dispute at Lionel Meeting," *New York Times*, Oct. 26, 1961, p. 47. "Embattled Boy Wonder; Roy Marcus Cohn," *New York Times*, Sept. 5, 1963, p. 25. "Ex-German's Plea in Behalf of a Colleague," *San Francisco Chronicle*, April 29, 1985, p. 7. "Space Scientist Admitted Role in Nazi Camp," *New York Times*, Oct. 6, 1985, p. 29. Wolfgang Saxon, "Arthur

Rudolph, 89, Developer of Rocket in First Apollo Flight, Dies" *New York Times*, Jan. 3, 1996, p. D18. "Portrait of Nazi Prompts Protest," *New York Times*, Oct. 26, 1993, p. A16.

5. Michael Neufeld, "Wernher von Braun, the SS, and Concentration Camp Labor: Questions of Moral, Political, and Criminal Responsibility," *German Studies Review*, Vol. 25, No. 1 (2002). Ben Franklin, "Von Braun Fights Alabama Racism," *New York Times*, June 14, 1965, p. 39. "Von Braun Recalls Sputnik 1 and Finds Five Areas of Challenge to Man in Space," *New York Times*, Sept. 24, 1967, p. 80. Nixon's budget cuts from http://history.nasa.gov/ SP-4012/vol4/ch1.htm SP-4012 NASA. John Noble Wilford, "Wernher von Braun Dies," *New York Times*, June 18, 1977, p. 49. Richard Evans, "The Sins of the Rocketeers," *New York Times Book Review*, Jan. 1, 1995, p. 2. Bernard Weinraub, "Huntsville's Joy Has a German Flavor," *New York Times*, July 25, 1969, p. 29. Bob Ward, *Dr. Space* (Annapolis: Naval Institute Press, 2005), provides a straightforward account of his life that dwells mainly his accomplishments. More critical is Michael Neufeld, *The Rocket and the Reich* (New York: Free Press, 1995).

6. Author interview with Bradford Whipple.

7. For Hynek, see the Web site of the J. Allen Hynek Center for UFO Studies, http://www.cufos.org/org.html. See also Joan Cook, "J. Allen Hynek, Astronomer and UFO Consultant, Dies," *New York Times*, May 1, 1986, p. D27.

8. Author interview with Roger Dobson.

9. Author interview with JoAnn Hardin Morgan. See also her official NASA biography at www.nasa.gov/centers/kennedy/about/biographies/morgan.html-31k; and see www.spaceref.com/news/viewpr.html?pid=12205-27k-.

10. Author interviews with Jay Barbree and Jo Ann Reisinger Barbree.

11. Author interview with Wickham Stivers Wilson.

12. A. C. Melissinos, "Nicholas C. Christofilos: His Contributions to Physics," CERN, European Organization for Nuclear Research, reports, *Cern*, 1995, pp. 1067–81. See Carol Altgelt, "The World's Largest Radio Station," www.oldradio.com/archives/jurassic/ELF.doc.

13. Author interview with Herbert York. See also Harry Kreisler, "Reminiscences from a Career in Science, National Security, and the University: A Conversation with Herbert York," Institute of International Studies, Berkeley, California, http://globetrotter.berkeley.edu/people/York/york88-con0.html-5k. http://www.er.doe.gov/fermi/html/Laureates/2000s/herberf.htm.

14. Author interview with Curtis Johnston.

INDEX

ABC, 96, 117, 121
AC Spark Plug, 67
Adams, Sherman, 102
Advanced Research Projects Agency
    (ARPA), 92, 139, 228
  SCORE project and, 229–30
Aeroballistics and Guidance Control
    Center, 157
Aerobee rocket, 18, 22
Aeromedical Research Laboratory, 214
Air Defense Missile Command, 82
Air Force, 126
Air Force, U.S., 23, 31, 53, 56–57, 61,
    69, 70, 75, 137, 159, 176, 189,
    194, 212, 223, 227, 229
  black bears experiment of, 91
  Blue Book Project of, 46
  manned flight and, 214–17
  MISS proposal of, 215–16
  moon shot by, 217–19
  Security Service Operations Center
    of, 26
  UFO phenomenon and, 82, 256

Air Force Association, 184
Albert I (monkey), 48
Albert II (monkey), 48
Albert III (monkey), 48
Albina (experimental dog), 86
Alsop, Joseph, 5
Alsop, Stewart, 5
"American Idealism in Action," 201–2
American Rocket Research Society,
    126
Antarctica, 21
Antigua, 163
Apollo-Soyuz, 257
Apt, Milburn, 215
Argus, exercise, 62–63
Argus, project, 203–8, 209, 259
  Explorer IV and, 205–6
Arkansas City Technological
    Organization of Rocketry and
    Aero Ballistics, 125
Armbrust, David, 19, 24–25, 26
Armed Forces Radio, 24
Armstrong, Neil, 215

Army, U.S., 31, 53, 55, 56, 58, 60, 63,
    77, 107, 189, 217
  NASA's bureaucratic conflict with,
    221–24
  satellite program of, 22–23, 44, 94,
    98, 105–6, 110, 130, 141, 213–14
  youth intern project of, 195–96
Army Ballistic Missile Agency
    (ABMA), 22, 30, 35, 69, 149–50,
    151, 155, 168, 169, 197, 223, 230
Army Corps of Engineers, 193, 213
Army Signal Corps, 229, 239–42
Army Transportation Corps, 213
Ascension Island, 199
Associated Press, 98, 111, 113, 198,
    236
Astronautics, 126
astronauts, 215, 232–33
Atlas Agenda missile, 260
Atlas rocket, 141–42, 176, 198–99,
    213–14, 230, 232
  Big Annie, 141
  New York Times on success of, 244–
    245
  10-B model of, 228–29, 236, 237,
    238, 240–42, 243, 244
Atomic Energy Commission, 137,
    204
atomic weapons, 2–3
  disaster novels of, 4
Australia, 44–45
Aviation Week, 93, 137, 236

B-47 bomber, 41
B-52 bomber, 41
Bahamas, 74
Baltimore Sun, 223
Barbree, Jay, 110–11, 121, 217–18
  later career of, 258
  as Mobile Mike, 192–95
  SCORE project and, 237

Barbree, Jo Ann Reisinger, 195, 258
Barr, Norman Lee, 233–35
Barr, Norman L., Jr., 233
Baruch, Bernard, 92
Baykonur space center, 86, 220
Belgium, 33
Belichick, Steve, 103
Bell Telephone, 67
Bell X-2 Starbuster, 215
Belyanka (experimental dog), 212
Bennett, Rawson, 40
Berlin blockade, 99
Bernays, Edward, 228
Bernds, Edward, 232
Beswick, Jerry, 116–17
Betts, Austin W., 58
Bide-A-Wee Home Association, 90
Big Annie (Atlas rocket), 141
Big Brother satellite, 93
Bioflight I, 234–35
Blagonravov, Anatoli, 29–30, 56,
    137
Blue Book, Project, 46
Bolshevik Revolution, 88
Bomarc missile, 74
bomber gap, 41
Boselli, Elizabeth, 99
Bracker, Milton, 198–99
Brazil, 74
Brewton, Charles, 95
Brezhnev, Leonid, 251
Bridges, Styles, 43
Brodsky, Jake, 66
Brucker, Wilbur M., 139, 164, 223
Buchanan, James, 138
Bulganin, Nikolai, 78
Bulgaria, 26
Bull Goose test vehicle, 72
Bumper 8 rocket, 72
Byelorussia, 2
Byrd, Richard, 21

Cahill, Larry, 17–18, 24
Cairnes, Dalton, 73, 197
California, University of, Berkeley, 61–62
California Institute of Technology, 22, 44
Cape Canaveral, Florida, 56, 65–76, 142–43
culture shock in, 68–69
errant and failed rockets in, 73–76
growth and development of, 63–67, 71
local economy of, 67–68
space agency workers in, 70–72
test range of, 72–74
Case Western University, 221
Cassini program, 252
CBS, 109, 120, 219
CBS News, 55
Central Intelligence Agency (CIA), 29
Corona satellite and, 63
Challenger disaster, 258
Checkers (Nixon's dog), 90
Chekunov, Boris, 13
Chief Designer, *see* Korolev, Sergei Pavlovich
China Lake, 48
"chins-up" speeches, 93
first, 96–97
second, 98–99
Christofilos, Nicholas, 61–62
Argus Project and, 203–8
electron shield hypothesis of, 108
later career of, 259
Christofilos effect, 204–6
Chrysler Corporation, 5, 59, 73
civil rights movement, 40
Clinton, Bill, 259
Close Encounters of the Third Kind (film), 82, 256

Cohn, Roy, 40, 253
Cold War, 4, 42, 52, 59, 83, 89, 109, 139, 150, 188, 192, 244, 246
arms race in, 4–5
McCarthyism and, 40
media and, 4–5
Sputnik I and, 14–15, 80
College News Conference, 214
Collier's, 34
Colombia, 256
Columbine III, 99, 135
Comer, Mason, 175
Communist Party, Soviet, 11, 140
Congress, U.S., 21, 40, 58, 59, 109, 137, 150–51, 168, 172, 173, 189, 211, 212, 222, 225
Congressional Research Service, 43
Conover, Donald, 232
Convair corporation, 59, 67, 228–29, 232, 236, 237–38
Corona surveillance satellite, 63, 247
Corporal missile, 97
Crossfield, Scott, 215
Cuban Missile Crisis, 247
Cunningham, Chester B., 28–29, 78, 80, 118

Daily Express, 123, 166
Daily Herald, 123
Daily Mirror, 123
Darwin, Charles, 20
Day, Doris, 232
Deal satellite, 60–61, 163–64
Debus, Kurt, 155–56, 157, 160–62, 165, 196
Deep Freeze, Operation, 17
Defense Department, U.S., 5, 9, 23, 94, 106, 137, 139, 141, 172, 183, 212, 213, 216, 223, 224, 243
failed rockets reports of, 112–13

Defense Department, U.S., *(cont.)*
Research Projects Agency of, *see*
Advanced Research Projects
Agency
de Havilland Comet, 14
DelVecchio, Philip, 45
DelVecchio, Robert, 45
Democratic Party, U.S., 40, 94, 96–97,
123, 131, 138, 210
Disney, Walt, 34–35, 173, 198, 213
Disneyland, 34
Dobson, Marilyn Smith, 257
Dobson, Roger, 65–66, 68, 257
Doherty, Tom, 159
Dornberger, Walter, 32
Douglas Aircraft, 76
Dowling, John, 45
Dulles, John Foster, 9, 101
Dyson, Freeman, 208

Easley, Johnny, 126
82nd Airborne Division, U.S., 31,
35
88 club, 236–37
Eisenhower, Dwight D., 3, 8, 9, 35, 39,
43, 55, 99–102, 135, 154, 155, 167,
172, 207, 212, 225, 227–28, 247
"chins up" speeches of, 93, 96–99
Democrats' criticism of, 92–96
Fulbright's criticism of, 138–39
Gaither Commission as seen by, 92–
93
integration issue and, 100, 246
Medaris's view of, 69–70
minor stroke of, 101–3, 135–36, 141
NASA-Army conflict and, 222–23,
224
NASA creation announced by, 188–
189
Open Skies proposal of, 42
in reaction to *Sputnik I*, 29, 40
in reaction to *Sputnik II*, 91–92, 137
satellite program and, 23–24
science advisor appointed by, 97–98
SCORE project and, 228, 229, 242–
243
State of the Union Address of, 173, 246
Suez crisis and, 100
"surrender" issue and, 210–11
U-2 project and, 41
Vanguard project and, 176–77
Vanguard TV-3 launch and, 116–17,
118, 123
York's conversation with, 92
Eisenhower, Mamie, 102, 117, 167
Eisenhower administration, 23, 24, 35,
124, 135, 152, 213
Senate Preparedness hearings and,
108–10
Sputnik I and, 29, 40–41, 43, 44
Engineering of Consent, The (Bernays),
228
environmental movement, 250
Escher, William, 175
Explorer I satellite, 163–66, 168, 172,
176, 179, 182, 192, 210
see also Juno/*Explorer I* launch
Explorer II satellite, 186–88, 210
Explorer III satellite, 186, 187, 210
Explorer IV satellite, 197
Argus project and, 205–6
Explorer V satellite, 230
Expo '58 world's fair, 200–202

Fairchild, 245
Fala (FDR's dog), 90
Farrell, Donald, 184–85, 214
Faubus, Orville, 100
Fermi Award, 259
Flatbush Society for the Fabrication
and Ignition of Pyrotechnic
Projectiles, 126

Ford, Gerald, 124
Ford, Henry, 6
Ford Fairlane, 6, 14
Ford Foundation, 92
France, 30, 136, 155
Frankel, Charles, 182–83
Freud, Sigmund, 228
Fulbright, J. William, 138–39
Furlong, William, 55

Gabor, Zsa Zsa, 232
Gagarin, Yuri, 250, 251
Gaither, H. R., 92
Gaither Committee, 92–93
Galbraith, John Kenneth, 245
Galileo program, 252
Galloway, Eileen, 43
Gavin, James "Slim Jim," 31, 35–36, 224
  resignation of, 139
Gellhorn, Martha, 35
Gemini program, 257
General Electric (GE), 57, 59, 125, 147, 228, 238, 244, 259
Germany, Federal Republic of (West Germany), 200
Germany, Nazi, 32, 99, 173
Glacier, USS, 17–19, 20, 24, 25–26, 60, 105
Glenn, John, 193, 244
Glennan, T. Keith, 221–24, 227, 253
Glenn L. Martin Company, 57–58, 59, 125, 144, 177, 181
Gniewk, John, 24
Goddard, Robert, 78
Gold, Herbert, 218
Goodpaster, A. J., 93, 102
Gordo (squirrel monkey), 233–36
Gorneau, Arnold, 126, 128
Gould, Jack, 209
Grand Central Rocket, 59

Gray, Norris, 147
Gray, Robert, 116, 175
Graybiel, Ashton, 234
Great Britain, 30, 32, 74, 90
Greensfield, Terry, 160
Griffith, C. E., 197
Grine, Kenneth, 112, 113, 154, 159, 165, 216, 217, 237

Hagen, John P., 30, 56–58, 118, 119, 122, 124, 145, 148, 176, 177, 180
Hagerty, James, 40, 118, 122, 136, 167, 210
Hansen, Anke, 200, 202–3, 255–56
Hardin, Donald, 68, 69
Hardin, JoAnn, 68, 69, 162, 196–97
  later career of, 257–58
Hardin, LaVerne, 68–70, 196
Harris, Gordon, 153
Haugland, Vern, 111
Hays, Wayne, 97
Hembree, Billy, 126
Hemingway, Ernest, 35
Hibbs, Al, 162–63, 165
Himmler, Heinrich, 32
Hinrichs, John, 223
Hitler, Adolf, 32, 39, 99
Hogjaw (radioman), 26–27, 200
Homer, 90
House of Representatives, U.S., 222
Human Decelerator, 49
Humphrey, Hubert, 96, 123
Huntsville, Ala., 149–51, 253
  Juno/*Explorer* celebration in, 168–69
Huntsville Times, 90, 151
Hynek, J. Allen, 45–46, 80–81
  later career of, 256

Institute on Global Conflict and Cooperation, 260

intercontinental ballistic missile
(ICBM), 5, 53, 97, 136, 176, 189,
211, 214, 242, 245, 247
first successful flight of, 88
U.S.'s first, 141–42
intermediate range ballistic missile
(IRBM), 75, 136, 194
International Geophysical Year (IGY),
20, 22, 25, 29, 45, 77, 81, 174,
185
Iowa, University of, 18, 60, 252
Isetta automobile, 193
isolationism, 52
I've Got a Secret (television show), 199

Jackson, Henry, 43, 123
Japan, 3
Jerry (Atlas missile worker), 142
Jet-Age Conference, 184
Jet Propulsion, 126
Jet Propulsion Laboratory (JPL), 22,
60, 94, 105, 107, 152, 162, 164,
165, 166, 222, 224, 231
Johnson, Lyndon B., 42–43, 94–95,
96, 123, 138, 168, 173, 209, 247
Senate Preparedness hearings of,
108–10, 129–30, 131, 135
Johnson, Roy, 228–29, 244
Johnston, Curtis, 237–42, 260
Joubert, Philip, 132
Jung, C. G., 82–83
Junior Chamber of Commerce, U.S.,
50
Juno/*Explorer I* launch, 142, 149–69,
174, 182, 195
confirmation of success of, 162–64,
166
delays in, 154–55, 158
Eisenhower's reaction to, 167
film on, 169
Huntsville celebration of, 168–69

jet stream problem and, 154–55
launch in, 160–62
Medaris and, 149, 151–55, 157–60
media and, 142, 152, 154, 157, 158–
159, 160, 162, 164–65, 168
orbit of, 166
praise and credit for, 167–69
press conference on, 167
public relations and, 151–52, 153
radio signal of, 166
rocket preparation in, 156–57
rocket stages of, 152–53, 162
Soviet Union and, 158
Van Allen and, 152, 153, 164, 165–
166, 167
von Braun and, 164, 165, 167, 168
Juno II missile, 230–31
Jupiter (planet), 244
Jupiter rocket, 93, 110, 111, 112, 131,
141, 186, 213, 230, 234
A type, 74–75, 196
C type, 75, 98, 151, 172
Justice Department, U.S., 253

Kaplow, Herbert, 218
Karpiscak, Paul, 119
KC-135 airplane, 99
Kearns, Doris, 42
Kennedy, John F., 95, 96, 250
Kennedy Space Center, 258
Kerouac, Jack, 39
Khrushchev, Nikita, 5, 42, 78–79, 99,
188, 220, 246
named "Man of the Year," 140
Reston's interview of, 52–54
satellite propaganda and power of,
88–89
silo idea of, 220
Sputnik and, 14, 47
Killian, James R., 41, 97–98, 109, 172,
177, 204, 222

Kimball, Ward, 34
Kimpton, Lawrence, 139–40
Kinchloe, Ivan, 215
King, Jack, 198
Kinsey, Alfred, 39
Kitzmiller, Donald, 259
Korean War, 2
Korolev, Sergei Pavlovich, 7–14, 188,
    212, 220
  background of, 8
  later career of, 251–52
  moon probe and, 246–47
  Sputnik I design and development
    and, 9–12
  at *Sputnik I* launch, 13–14
  Sputnik II and, 46, 51–52, 78–79, 86,
    88
Kudryavka, *see* Laika

Laika (experimental dog), 85–87, 94,
    182, 188, 197, 201
  description of, 85, 87
  in *New York Times*, 99
  nickname of, 86–87
  in orbit, 90–91
  in rocket launch, 87–88
  Talese on, 90
Lark rocket, 72
LeMay, Curtis, 3, 216
  in nonstop record flight, 99
Lemnitzer, Lyman, 107
Lenin, V. I., 201
Lenin Prize, 251
Leonov, Alexei, 251
Levitt, Israel, 44
Levittown, 2
Ley, Willy, 32, 90, 125
Life, 62, 163, 210, 235–36
Lionel Corporation, 253
Little, Gordon, 45
Lockheed Corporation, 41

Los Angeles Times, 251
Luce, Clare Booth, 140
Luce, Henry, 140, 210
Ludwig, George, 60, 166, 186, 188
Luna I, 246–47
Lundquist, Charles, 165–66

McCall, Gene, 111
McCarthy, Joseph, 40, 253
McElroy, Neil H., 35, 36–37, 93–94,
    98, 245
McGraw, Donald, 93
McIlwaine, Carl, 186–87
McNabb, G. G., 142
Making Weapons, Talking Peace
    (York), 259–60
Malina, Frank, 94
Malinovksy, Rodion, 251
Maloy, Travis, 237–40, 242
Manhattan Project, 59, 70, 209
Mao Tse-Tung, 47, 78, 89
Mariner program, 252
Mars, 129, 244, 255
Mars and Beyond (television show),
    117
Marshall Space Flight Center, 254
Martin, Ronald, 82
Massachusetts Institute of Technology,
    41, 97, 201
Mathison, Charles, 197
Mazur, Daniel, 57–58, 116, 175, 181
Mead, Margaret, 83
Mechta, see Luna I
Medaris, J. Bruce, 22, 56, 58, 60, 69,
    77, 97, 98, 106, 172, 173, 176,
    196, 213, 217, 222, 230
  background of, 30–31
  bureaucratic aggressiveness of, 167–
    168
  congressional testimony of, 129–32
  Eisenhower as viewed by, 69–70

Medaris, J. Bruce, *(cont.)*
 Juno/*Explorer I* launch and, 149,
  151–55, 157–60
 later career of, 253–54
 militant outlook of, 69–70
 NASA–Army conflict and, 223–24,
  227
 space agenda of, 129–31, 132
 Sputnik reaction of, 35–37, 46
media:
 Air Force's moonshot and, 217–19
 animal space flights and, 198–99
 chins-up speeches and, 96
 Cold War and, 4–5
 Juno/*Explorer I* launch and, 142,
  152, 154, 157, 158–59, 160, 162,
  164–65, 168
 NASA–Army conflict and, 223–24
 Senate Preparedness hearings and,
  109, 130
 Sputnik "crisis" and, 40, 44, 55, 59
 Vanguard project and, 76, 144–46
 Vanguard TV-3 test and, 110–14,
  120–21, 123, 124
Mercury (planet), 244
Mercury, Project, 224–25, 257, 260
Mia (mouse), 197
Mia 2 (mouse), 197
MiG-15 fighters, 2
Mills, C. Wright, 40
Minitrack network, 28, 78, 118
Mi-6 helicopter, 88
MISS (Man in Space Soonest),
  215–16
Missile 29, 35, 37, 107, 109, 130
 see also Juno-*Explorer I* launch
Mobile Mike, *see* Barbree, Jay
model rocketry, 125
Moon, 129, 194, 212, 213, 216–20,
  250, 254–55
 failed Air Force shot at, 217–19

 Soviet probe to, 246–47
Morgan, JoAnn Hardin, *see* Hardin,
  JoAnn
Morocco, King of, 101, 102
Moser, Robert, 156–57, 159–61
Motorola, 245
Mutual Assured Destruction (MAD),
  3, 136, 247
Mutual Broadcasting Network, 246

Nasser, Gamal Abdel, 100
National Academy of Sciences, 29, 56,
  137, 152, 167, 172, 187
National Advisory Committee for
  Aeronautics (NACA), 221–22
National Aeronautics and Space
  Administration (NASA), 227, 230,
  252, 253, 254, 255, 258, 259
 Army's bureaucratic conflict with,
  221–24
 creation of, 188–89, 220–21
 JPL acquired by, 222, 224
 von Braun's proposal of, 133
National Airlines, 258
National Defense Education Act, 211
National Education Association, 138
National Guard, U.S., 100
National Science Foundation, 138,
  167, 221
National Security Council, 116, 204
Navaho missile, 70, 111
Naval Research Laboratory, 22, 44,
  77–78, 144
Navy, U.S., 48, 53, 56, 69, 77, 107,
  172, 194, 204, 212–13
 satellite program of, 22–23, 35, 36,
  44, 106, 110, 141, 214
 School of Aviation Medicine and
  Research of, 233
NBC, 96, 121, 194, 258
 Air Force moonshot and, 217–19

Neilon, John, 76–77, 118, 174–76, 177, 180–81
Neufeld, Michael, 32–33, 254
New York Stock Exchange, 76
New York Times, 1, 11, 29, 44, 52, 54, 78, 90, 96, 99, 120, 167, 173, 198, 219, 224
  Missile 29 pictured in, 130
  on success of Atlas rocket, 244–45
Nikitin, Nikifor, 12
Nixon, Richard M., 90, 102, 255
  Cow Palace speech of, 211
Nobel Prize, 109
North American Aviation, 4, 67
North Atlantic Treaty Organization (NATO), 135–36, 138, 141
  first meeting of, 135–36, 138, 141
Northwestern University, 256
Norton Sound, USS, 204, 205

Odyssey (Homer), 90
"Old Number Seven," 8
O'Malley, Thomas, 193–94
101st Airborne Division, U.S., 100
On the Beach (Shute), 4
Open Skies, 41–42
Operation Deep Freeze, 17
Oppenheimer, Robert, 62
Orange (hydrogen bomb), 205
Orion, Project, 208
Oscar (black bear), 91

Peace Corps, 255–56
Pearson, Drew, 209
Pentagon, *see* Defense Department, U.S.
Pershing rocket, 151
Peru, 27–28, 78, 80, 118
Pestraya (experimental dog), 212
Pickering, William, 60, 105, 153, 164, 165, 166, 167, 231, 232

Pie Truck, 218
Pinson, V. L., Jr., 196
Pioneer Program, 252
Pioneer 0 satellite, 217
Pioneer I satellite, 219, 220
Pioneer II, 231, 232
Pioneer III, 231
Pokrovsky, Alexei, 85–86
Polaris missile, 194
"Polar Years," 20
Porter, Sylvia, 111
Poulter, Thomas, 21
Power Elite, The (Mills), 40
Pravda, 14, 47, 251
Preliminary Design of an Experimental World Circling Spaceship, 216
President's Science Advisory Committee (PSAC), 172, 204
Presley, Elvis, 40
Press-Citizen, 116
Procter & Gamble, 35
Project Blue Book, 46
Project Mercury, 224–25, 229, 260
Project Orion, 208
proximity fuse, 21, 59

Quarles, Donald, 223
Queen of Outer Space (film), 232
quiz show scandal, 209

R-7 rocket, 8–10, 11, 12, 13, 30, 53, 79, 86, 87, 132, 136, 188, 213, 220
R-11 missiles, 10
R-12 rocket, 10
Rabi, Isidor, 62
Ramo, Simon, 214
Ramo-Woolridge company, 214
RAND Corporation, 216
Raytheon, 245
RCA, 26, 44, 80, 130

Reasoner, Harry, 120–21
Redstone Arsenal, 30, 31, 35–36, 56–57, 63, 68, 97–98, 106, 107, 134, 150–51, 156, 168
 German Colony of, 35
Redstone rocket, 4–5, 22, 23, 25, 55, 60, 69, 74–75, 131, 213, 234
Reedy, George, 95
Republican National Committee, 167
Republican Party, U.S., 40, 95, 96, 97, 138, 211
Reston, James "Scotty," 1–2, 6–7, 11, 96
 Khrushchev interviewed by, 52–54
Reston, Richard, 1, 6–7, 14–15
 later career of, 250–51
Reston, Sally, 1, 6
Reuters, 198
Riddell, Mary, 90–91
Rocket and the Reich, The (Neufeld), 254
Rocket City, 35
Rocketdyne, 59
rockoon, 17–20, 22, 25, 185
Roosevelt, Franklin D., 5, 90
Rostow, Walt, 201
Royal Air Force, 132
Royal Astronomical Society, 252
Royal Swedish Academy of Sciences, 252
Rubenstein, Paul, 120
Rudolph, Arthur, 253–54
Russell, Anna, 172

St. Louis Globe Democrat, 223
satellite program, *see* space race
Saturday Evening Post, 5, 20
Saturn (planet), 208, 252
Saturn rocket, 225, 254
Saturn V, 255
Saucedo, Pedro, 82

Saudi Arabia, 260
School of Aviation Medicine, 214
SCORE (Signal Communication by Orbiting Relay Equipment) Project, 228, 236–44, 245, 259, 260
 ARPA and, 229–30
 88 club and, 236, 237
 Eisenhower's voice broadcast from, 242–43
 fueling process in, 240
 launch of, 242–43
 nosecone in, 238–39
 scientific instruments in, 239–40
 secrecy of, 236–37
Scoville, Herbert, Jr., 29
Sedov, Leonid, 44, 107
See It Now (television program), 109
semiconductor industry, 245–46
Senate, U.S., 42–43, 139, 222
Senate Preparedness Subcommittee hearings, 108–10, 129–35
 Medaris's testimony in, 129–32
 Teller's testimony in, 108–9
 von Braun's testimony in, 129, 132–35
Sendler, Karl, 196
Serling, Rod, 192
Sevareid, Eric, 55
Shepard, Alan, 250
Shute, Nevil, 4
Siegel, Gerald, 42
Simons, David, 49–51, 214, 216
Skunk Works, 41
Skylab, 257
Small, John, 60
Smithsonian Astrophysical Laboratory, 45
Snark rocket, 72, 74
Snow Cruiser, 21, 25
Snyder, Howard, 102, 117, 124

Soviet Union, 1–11, 35, 51, 55, 63, 77, 78, 93, 97, 99, 107, 109, 135, 168, 188, 214, 220, 225, 242, 244, 245
  animals recovered from space by, 211–12
  Argus project and, 206, 207
  bomber gap and, 41
  Chelyabinsk disaster in, 140
  collapse of, 247
  Expo '58 pavilion of, 200–202
  grand technological claims of, 137–138
  hydrogen bomb of, 3
  Juno launch and, 158
  military technology coups of, 88–89
  moon probe of, 246–47
  Open Skies proposal rejected by, 41–42
  Reston's visit to, 2–3, 6–11
  space effort motto of, 191–92
  U.S. contrasted with, 2, 5–6, 11
space race:
  animal experiments in, 48, 183–84, 197–200, 212, 233–36
  antisatellite experiment in, 62–63
  assessment of, 249–50
  astronaut program and, 232–33
  bureaucratic infighting and, 22–23
  congressional funding of, 150–51
  cost of, 249
  firsts in, 191
  Hollywood and, 232
  manned flight goal in, 48–51, 182–188, 212, 214–17, 232–33
  model rocketry fad and, 125–28
  national defense and, 212
  as political issue, 138–40
  prestige and, 212
  PSAC and, 172–73
  rocket underground and, 60–61
  science education and, 209–11
  Sputnik I and onset of, 55–56
  U.S. economy and, 192, 245–46
  Van Allen radiation belts and, 185–186
  von Braun on, 173–74
  zero count riddle and, 186–87
  see also specific programs, events, and satellites
Space School, 231–32
Spain, 44
Speer, Albert, 32–33
Spielberg, Steven, 82, 256
Sputnik I, 86, 88, 100, 107, 137, 150, 172, 182, 188, 225, 247, 249, 250–51, 255
  Cold War and, 14–15, 80
  design and development of, 9–11
  early U.S. satellite experiments and, 60–64
  Eisenhower administration's reaction to, 40–41, 43, 44
  Eisenhower's reaction to, 29, 40
  exhibition models of, 201–2
  first visual observation of, 45
  follow-up to, *see Sputnik II*
  in Khrushchev–Reston interview, 52–53
  launch of, 13–14, 15
  Lyndon Johnson's reaction to, 42–43
  media and, 40, 44, 46, 55, 59
  one-year anniversary of, 47, 210–11
  onset of space race and, 55–56
  press conference on, 54
  as propaganda coup, 25, 27, 29, 41, 43–44
  radio signal of, 24–29, 30, 44–45, 80
  Redstone group's reaction to, 35–37
  secret message rumors and, 44–45
  UFO phenomenon and, 81–83, 191
  U.S. economy and, 192

Sputnik I, (cont.)
  world's reaction to, 14–15, 29–30,
    40–43, 46, 80–81, 97–98
Sputnik II, 47–48, 133, 148, 210
  design and development of, 78–79
  dog selected for, *see* Laika
  Eisenhower's reaction to, 91–92, 137
  launch of, 87–88
  orbit of, 87–88
  public reactions to, 97–98, 136–37
  as public relations coup, 89–91
  weight of, 79, 86
Sputnik III, 188–89, 210, 228
Stalin, Joseph, 7–8, 88, 251
Stapp, John, 49, 51, 183–84, 214, 216
Star Trek (television show), 212
State Commission, Soviet, 7, 11, 13
State Department, U.S., 52, 96
Stehling, Kurt, 119–20, 122–23, 142,
    143–44, 147, 153, 175, 178, 179–
    180, 181
Stennis, John, 133
Stephens, Ralph, 81
Stevenson, Adlai, 96
Stivers, Eugene, 258
Stivers, Wickham "Wickie," 111–12,
    113, 144–45, 164–65, 198, 217
  later career of, 258–59
Strategic Air Command (SAC), 3, 99
Stuhlinger, Ernst, 106, 107, 117
Suez Canal, 100
Sullivan, Walter, 29
Symington, W. Stuart, 43, 134, 189

Talese, Gay, 90
Tarawa, USS, 204
Tass (news agency), 14
Tate, Michael, 125–26
Teak (hydrogen bomb), 205
Teller, Edward, 62
  congressional testimony of, 108–9

Test Vehicle-3, *see* Vanguard TV-3 test
Texas Instruments, 245
Thor-Able rocket, 197, 217, 219
Thor rockets, 75–76, 111, 124, 141,
    143, 213, 218, 219
Three Stooges in Orbit, The (film), 232
Time, 4, 108, 154, 173
  Khrushchev named "Man of the
    Year" by, 140
  Van Allen named "Man of the Year"
    by, 252
Titan missile, 213–14
To Tell the Truth (television show), 199
Trinidad, 176
Trudeau, Arthur, 223
Truman, Harry S., 40, 96, 210
Tunnel of Love, The (film), 232
TU-114 airplane, 88
Turkey, 8
Twilight Zone, The (television show),
    192

U-2 spy plane, 41, 53
  first flight of, 47
UFO Experience, The (Hynek), 256
unemployment, 192
unidentified flying objects (UFOs), 46,
    256
  sightings of, 81–82, 191
United Nations, 101
United States:
  economy of, 192, 245–46
  Expo '58 pavilion of, 201–2
  in 1950s, 39–40
  Soviet Union contrasted with, 2, 5–6,
    11
  in test ban negotiations, 207
United States Weather Bureau, 126

V-2 rocket, 4–5, 32, 33–34, 47, 48, 94,
    134, 214, 232

Van Allen, Abigail, 171–72, 252
Van Allen, James, 17, 18–22, 56, 59–
61, 94, 116, 149, 176, 185, 204,
206, 210, 219
early career of, 21–22
Juno/*Explorer I* launch and, 152,
153, 164, 165–66, 167
later career of, 252
as "Man of the Year," 252
proximity fuse project and, 21, 59
radiation belts discovery and, 185–
187, 197, 231
Sputnik observed by, 24–26
at White House dinner, 171–72
Vance, Cyrus, 130
Van Doren, Charles, 209
Vanguard, Project, 28, 40, 54, 56–59,
63, 76–78, 89, 90, 98, 106–7, 110,
153, 168
Air Force bureaucracy and, 76–77
Eisenhower's commitment to, 77–78
failures and delays in, 57–58, 142–48
first test flight of, 57
funding of, 58–59
TV-3 test of, *see* Vanguard TV-3 test
Vanguard rocket, 23–24, 25, 27–28,
29, 55, 106, 172, 174, 210
Vanguard TV-3 test, 78, 110–25, 129,
142, 154, 158, 192, 195
back-up launches and, 174–75
description of, 114–15
Eisenhower and, 116–17, 118, 123,
176–77
failure of, 121–23, 130
first countdown in, 115–16
media and, 110–14, 120–21, 123,
124
postmortem on, 123–25
second countdown in, 117–20
Vanguard TV-4 rocket, 178–82
success of, 178–82

Venezuela, 74
Venus, 129, 244
Vietnam War, 251, 255
Viking rocket, 22, 23, 77
von Braun, Wernher, 30, 56, 58, 60,
69, 70, 77, 90, 107–8, 117, 148,
153, 176, 188, 213, 230, 232, 233,
244, 259
background of, 31–35
Big Brother satellite idea of, 93
congressional testimony of, 129,
132–35
Juno/*Explorer I* launch and, 164,
165, 167, 168
later career of, 254–55
NASA and, 133, 222, 223, 227
Nazi past of, 31–34, 173, 254
personality of, 106
popularity of, 34, 173
satellite program and, 35–36
on Soviet space achievement, 94
space agenda of, 98, 129–30, 131,
134, 213, 254–55
on space race, 173–74
U.S. rocket program and, 34–35
V-2 rocket and, 32–34
at White House dinner, 172
Von Fremd, Charles, 120
Von Fremd, Virginia, 120–21
von Karman effect, 114–15
von Quistorp, Maria, 34
Voyager program, 252

WALB, 110
Wallace, Mike, 114
Walsh, J. Paul, 118, 122
War and Peace in the Space Age
(Gavin), 224
War Department, U.S., 34
Washington, USS, 21
Watson, Mark, 223

Weisl, Edwin L., 56, 133
WEZY, 193
WFGA, 218
What's My Line (television show), 199
Whipple, Bradford, 26–27
  in Expo '58 escapade, 200–203, 255
  later career of, 255–56
Whipple, Fred Lawrence, 45, 80
Wickie Mouse, 198–99, 212
Widmark, Richard, 232
Wiley, Alexander, 40
Williams, G. Mennen, 39
Wilson, Charles E., 23, 36, 114, 168–169
Wilson, Stephen, 18–19, 20, 24, 26
Wilson, Woodrow, 92

World War II, 2, 3, 21, 22, 31, 52, 59, 94, 115, 135, 136, 173, 224
WRCA, 130

X *Minus 80 Days* (film), 169

Yates, Donald, 112, 116, 144, 153–54, 155, 157, 158, 236–37
York, Herbert, 29, 62, 92, 124, 139–40, 177, 207, 208, 224–25, 227–28, 244
  Eisenhower's conversation with, 92
  later career of, 259–60
  on SCORE project, 229

Zorro (television show), 96

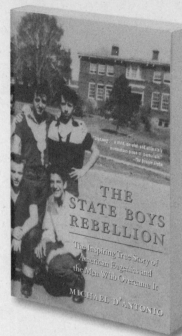

MICHAEL D'ANTONIO

"Thorough and highly readable . . . It is the great charm of D'Antonio's book that he will not plunk entirely for one judgment or the other. It's the man he's after, not the god."
—Benjamin Cheever, *The New York Times Book Review*

"A valuable addition to the literature of American business and philanthropy."
—Jonathan Yardley, *The Washington Post Book World*

"A charming and absorbing account of one of American capitalism's eccentric visionaries whose business success was matched by his devotion to social improvement . . . an intriguing portrait of a true American individualist and his time."
—Anthony Day, *Los Angeles Times Book Review*

"A richly detailed biography of the founder and an absorbing history of the Hershey company."
—Hardy Green, *BusinessWeek*